ADSORPTION

ADSORPTION

J. OŚCIK

Professor of Chemistry
Maria Curie-Skłodowska University
Lublin

Translation Editor:
I. L. COOPER
School of Chemistry
University of Newcastle upon Tyne

ELLIS HORWOOD LIMITED
Publishers · Chichester

Halsted Press: a division of
JOHN WILEY & SONS
New York · Chichester · Brisbane · Toronto

English Edition first published in 1982 in coedition between
ELLIS HORWOOD LIMITED
Market Cross House, Cooper Street, Chichester, West Sussex, PO19 1EB, England
and
PWN – POLISH SCIENTIFIC PUBLISHERS
Warsaw, Poland.
The publisher's colophon is reproduced from James Gillison's drawing of the ancient Market Cross, Chichester.
Translated by *Krzysztof Radziwill* from Polish edition *Adsorpcja*,
Państwowe Wydawnictwo Naukowe, Warszawa 1979
Distributors:
Australia, New Zealand, South-east Asia:
Jacaranda-Wiley Ltd., Jacaranda Press,
JOHN WILEY & SONS INC.,
G.P.O. Box 859, Brisbane, Queensland 40001, Australia.
Canada:
JOHN WILEY & SONS CANADA LIMITED
22 Worcester Road, Rexdale, Ontario, Canada.
Europe, Africa:
JOHN WILEY & SONS LIMITED
Baffins Lane, Chichester, West Sussex, England.
Albania, Bulgaria, Chinese People's Republic, Cuba, Czechoslovakia, German Democratic Republic, Hungary, Korean People's Democratic Republic, Mongolia, Poland, Romania, Vietnam, the U.S.S.R., Yugoslavia:
ARS POLONA – Foreign Trade Enterprise
Krakowskie Przedmieście 7, 00-068 Warszawa, Poland.
North and South America and the rest of the world:
Halsted Press: a division of
JOHN WILEY & SONS
605 Third Avenue, New York, N.Y. 10016, U.S.A.
British Library Cataloguing in Publication Data
Ościk, J.
Adsorption. – (Ellis Horwood Series in Physical Chemistry)
1. Adsorption
I. Title II. Adsorpcja. *English*
541.3'453 QD547
Library of Congress Card No. 81-6462 AACR2
ISBN 0-85312-166-4 (Ellis Horwood Ltd., Publishers)
ISBN 0-470-27218-X (Halsted Press)

Printed in Poland

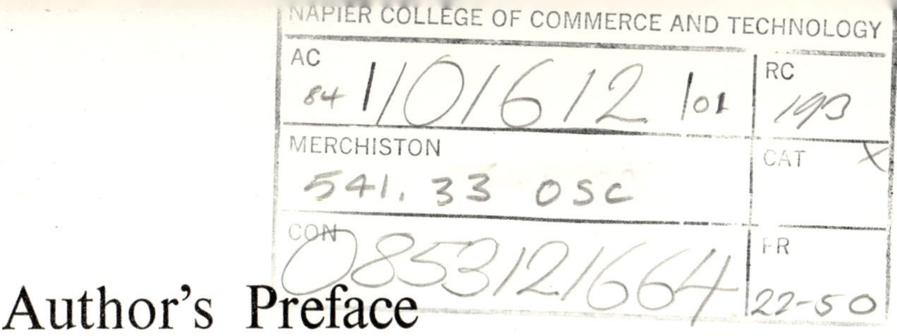

Author's Preface

During my study of adsorption processes over many years, it has often struck me, that there is no monograph which uniformly treats processes occurring at all kinds of interfaces. Also, student's textbooks in physical chemistry devote little space to the problems of adsorption, whereas adsorption is increasingly attracting the interest of wide circles of scientists whose investigations are related to heterogeneous systems. In such common systems the role of surface phenomena is often dominant, and adsorption is one of the principal processes. The growing interest in adsorption is also due to the rapid development of adsorption (gas and liquid) chromatography—now a common analytical method.

The above arguments have revealed the necessity for a book which would acquaint large numbers of research workers and advanced students in natural sciences and technology with the foundations of adsorption processes. In undertaking the preparation of this book I have presented these processes comprehensively, putting greatest emphasis on experimental and theoretical aspects of the described phenomena as well as on their thermodynamic features.

The terminology and notation used in the present edition have been based on the IUPAC proposals (*Manual of Definitions, Terminology and Symbols in Colloid and Surface Chemistry, Pure and Appl. Chem.*, **31**, 577 (1972)).

I wish to thank Doctors Andrzej Dąbrowski and Mieczysław Jaroniec for their kind help in the preparation of those sections of my book devoted to adsorption on heterogeneous adsorbent surfaces.

Jarosław Ościk

Contents

List of Symbols

a	activity of a substance in the bulk phase
a	quantity of substance adsorbed by 1 g of adsorbent
a_i	adsorbent surface area covered by 1 mole of adsorbate i
a_t	quantity of substance adsorbed by 1 g of adsorbent after time t
a	specific capillary activity, constant in Szyszkowski's equation
a_m	quantity of substance adsorbed in the monolayer (per 1 g of adsorbent) or monolayer capacity
a_{Mi}	ion activity in ion exchanger
A	surface area of interface
A	adsorbent surface area
A	surface area occupied by a molecule at the solution/gas (vapour) interface at surface pressure π
A_0	least surface area occupied by a molecule at the solution/gas (vapour) interface
b	constant in Szyszkowski's equation
B	constant in Szyszkowski's equation
B_i	constant in the general exponential adsorption isotherm equation
c_i	molar concentration of substance i in the bulk phase (solution) at equilibrium (after adsorption) or equilibrium concentration
c_i^0	molar concentration of substance i in the bulk phase (solution) prior to adsorption
c_i^g	molar concentration of substance i in the gaseous phase
c_n	molar concentration of saturated solution
c_i^s	molar concentration of substance i in surface (interfacial) layer
$c_{m,i}^s$	molar concentration of substance i in surface layer at complete coverage of the adsorbent surface
C	constant in the BET equation
C	quantity defined in equation (4.14) expressing the ratio of activity coefficients of a binary solution in the surface and bulk layers
C_c	heat capacity of calorimeter and adsorbent
C_{pm}	molar heat capacity of gas in the bulk phase
C_{pm}^s	molar heat capacity of gas in the surface layer
d	diameter of globules in corpuscular adsorbent structure
E	surface energy of adsorbent
f_i and f_i^s	activity coefficients of substance i in the bulk and surface phases
F	Helmholtz free energy

F^s	Helmholtz free energy of surface layer
F^σ	Helmholtz free energy of surface excess amount
G	Gibbs free energy
$\mathscr{G}^s, \hat{G}^s, G^s$	Gibbs free energy of surface layer (of adsorbed gas)
$\bar{G}^s$	differential Gibbs free energy of surface layer
$\bar{G}^s_m$	molar differential Gibbs free energy of surface layer (of adsorbed gas)
$\Delta_a G^s_m$	molar differential Gibbs free energy of gas adsorption
$\Delta_a G^s_{m,i}$	molar differential Gibbs free energy of adsorption of substance i from a solution
$\Delta_a G^s_i$	differential Gibbs free energy of adsorption of substance i
$\Delta G_{1,2}$	total change of Gibbs free energy of a system on successive addition of solution components 1 and 2
ΔG_l	change of Gibbs free energy connected with an equilibrium solution or Gibbs free energy of mixing of an equilibrium solution
ΔG^0_l	Gibbs free energy of mixing of the initial solution (prior to adsorption)
$\Delta_w G$	Gibbs free energy of wetting
$\Delta_w G^0_i$	Gibbs free energy of wetting with pure substance i
H	enthalpy of the bulk phase
H_m	molar enthalpy of the bulk phase
$\bar{H}_m, \bar{H}^s_m$	differential molar enthalpy of the bulk phase (gas in the bulk phase) and of the surface phase (adsorbed gas)
$\mathscr{H}^s, \hat{H}^s, H^s$	enthalpy of surface layer (of adsorbed gas)
$\Delta_a H^s_m$	differential molar enthalpy of gas adsorption or equilibrium adsorption heat
$\Delta_a \mathscr{H}^s_m$	differential molar enthalpy of gas adsorption
$\Delta_a H^s_{m,i}$	differential molar enthalpy of adsorption of component i from solution
$\Delta H_{1,2}$	change of enthalpy on successive addition to the adsorbent of solution components 1 and 2
ΔH_l	enthalpy of formation of equilibrium solution
ΔH^0_l	enthalpy of formation of initial solution (prior to adsorption)
$\Delta_w H$	enthalpy of wetting the adsorbent surface with solution or enthalpy of immersion of the adsorbent in solution
$\Delta_w H^0_i$	enthalpy of wetting the adsorbent surface with pure substance i
ΔH^s_l	enthalpy of formation of surface solution
ΔH^L_m	differential molar condensation enthalpy
$h=p/p_0$	relative gas (vapour) pressure
k	constant in the Langmuir and many other equations
k_L	condensation constant
k_a	gas adsorption rate constant
$K, K_{a,p}, K_{a^s,p}$	Henry's constant
K_1	equilibrium constant for adsorption from solution 1+2 in an ideal system
K_2	constant in the Hill and de Boer adsorption isotherm equation
K_n	constant in the Kiselev adsorption isotherm equation related to the

horizontal interaction in the surface layer

l constant in the Everett equation

l^s thickness of the adsorption (surface) layer

L molar heat of condensation

m mass of adsorbent

m constant in the Everett equation

$\bar{M}$ molecular mass

M average molecular mass of solution

n total number of moles in the bulk phase at equilibrium

n^0 total number of moles in the bulk phase prior to adsorption

n^s total number of moles in the surface layer or capacity of the surface layer (phase)

n_i number of moles of component i in the bulk phase at equilibrium

n_i^s number of moles of componont i in the surface layer at equilibrium

$n_{m,i}^s$ number of moles of component i in the surface layer at full coverage of the adsorbent surface

n^σ total surface excess amount

n_i^σ surface excess amount of component i or Gibbs adsorption of component i

$n_i^{\sigma(1)}$ relative adsorption of component i with regard to component 1

$n_i^{\sigma(n)}$ reduced adsorption of component i

$n_i^{\sigma(v)}, n_i^{\sigma(m)}$ surface excess amounts as defined by Guggenheim and Adam

$n_{i,1}^{\sigma(n)}(x_i)$ surface excess amount of component i on the homogeneous patch of the adsorbent surface or local excess adsorption isotherm of component i

N Avogadro's number

p outer gas pressure or equilibrium gas pressure

p_c saturated vapour pressure over a cylindrical liquid meniscus

p_i partial pressure of component i

p_{sph} saturated vapour pressure over a spherical liquid meniscus

p_0 saturated vapour pressure over a flat liquid meniscus

P_i coefficients in the Guggenheim and Adam equation

q_a differential molar adiabatic heat of adsorption

q_d differential molar heat of adsorption

q_{st} differential molar isosteric heat of adsorption

r capillary radius

r_c radius of cylindrical liquid meniscus

r_{sph} radius of spherical liquid meniscus

r_1 radius of curvature of the meniscus

r_T capillary radius calculated from the Kelvin equation

R gas constant

s specific surface area

s^s new surface formation entropy

S_m molar entropy of the bulk phase (gas)

$\bar{S}_m$ differential molar entropy of the bulk phase (gas)

$\bar{S}_m^s$ differential molar entropy of the surface layer (adsorbed gaseous layer)

$\Delta_a S^s_{m,A}$, $\Delta_a S^s_{m,\gamma}$ differential molar gas adsorption entropy
T temperature of measurement
U_m molar internal energy of the bulk phase
U^s_m molar internal energy of the surface phase (adsorbed gas)
$\bar{U}^s_m$ differential molar internal energy of the adsorbed gas
$\Delta_a U^s_m$ differential molar internal energy of the gas adsorption
V volume of the system
V_a volume of the adsorbent
V^l volume of the liquid phase
V^g volume of the gaseous phase
V_m molar volume
$\bar{V}_m$ average molar volume of the solution
V^s_m molar volume of the adsorbed gas
$\bar{V}^s_m$ differential molar volume of the adsorbed gas
V^s volume of the surface layer
V^s_0 limiting volume of the surface layer for $\varepsilon=0$
V_p volume of pores
W^s work required to transfer 1 mole of substance from the bulk phase to the surface layer
$W^0_{w,i}$ work of wetting with pure component i
x_i, x^s_i mole fractions of component i in the bulk phase and in the surface phase at adsorption equilibrium
x^0_i molar fraction of component i in the bulk phase prior to adsorption
α distribution coefficient or distribution function
α^s amount of substance in the surface layer per unit surface area of adsorbent or surface concentration
α^s_0 surface concentration of free active sites on the adsorbent surface
α^s_m surface concentration at complete coverage of the adsorbent surface
β coefficient of surface displacement
β_a convergence (affinity) coefficient of the characteristic curves
β_p constant in the Prigogine and Defay equation
γ surface (interfacial) tension
Γ surface excess concentration
ε adsorption potential or adsorption energy (free adsorption energy)
θ wetting or limiting angle
θ, $\theta_t(p)$, $\theta_t(x)$ coverage of the adsorbent surface with adsorbate molecules
$\theta_l(p, \varepsilon)$ degree of coverage of a homogeneous patch or the adsorbent surface or local gas adsorption isotherm
$\theta_{i,l}(x, \varepsilon)$ degree of coverage of a homogeneous patch of the adsorbent surface with component i of the solution or local adsorption isotherm of component i of a solution
μ_i chemical potential of component i of a solution
$\mu_i^{\ominus}$ standard chemical potential of component i
μ_i^0 chemical potential of pure component i
μ_i^s chemical potential of component i in surface layer
$\mu_i^{\ominus,s}$ standard chemical potential of component i in the surface layer
$\mu_i^{0,s}$ chemical potential of pure component i in the surface layer

$\Delta_a\mu_i^\ominus$	standard chemical potential of adsorption of component *i*
μ, μ^s, μ^σ	chemical potentials of gas in the bulk phase and surface layer, and the chemical potential of the surface excess amount of gas
ν	constant in the Everett equation
π	surface pressure
ρ	density
τ_1, τ_L	residence times of adsorbate molecules in the first and further layers of the surface phase
φ_i	volume fraction of component *i* in the bulk phase
$\chi(\varepsilon)$	adsorption energy distribution function
ω_0	surface area per adsorbed molecule
ω_m	surface area occupied by one adsorbate molecule in the monolayer or sitting surface

Chapter 1

Introduction

1.1 General

Many physical and chemical processes occur at the boundary between two phases, while others are initiated at that interface. An understanding of phenomena occurring at such boundary surfaces is therefore often essential for explaining the mechanism of, for example, dissolution and crystallization, electrode processes, heterogeneous catalysis and phenomena related to the colloidal state.

The role of surface phenomena is frequently underestimated or overlooked, although surfaces play a significant role in all natural sciences. Detailed knowledge of surface phenomena is therefore indispensable for all concerned with these sciences.

Since adsorption is one of the fundamental surface phenomena, we shall concentrate in the present book on its description, the theories underlying it and on certain methods for its investigation.

1.2 General Concepts, Definitions and Classification of Adsorption Systems

Van der Waals or cohesive forces act between the molecules of all substances irrespective of their state of aggregation. Such forces are revealed only when the molecules are not more than several nanometres (1 nm= $=10^{-9}$ m) apart. The greatest distance between molecules at which cohesive forces still act is called the radius of molecular interaction. In the bulk of a phase, intermolecular cohesive forces are balanced. In solids, liquids or gases the atoms, ions or molecules in the interface (surface layer) are exposed to the action of unbalanced forces due to both phases, these resultant forces acting normal to the phase boundary. Hence, for any phase:

(i) The surface molecules or atoms are in a different energy state compared with those in the bulk phase. This additional energy, known as the surface energy, imparts to the surface region certain distinct features which differ significantly from those of the bulk regions of the phase.

(ii) The total internal energy of a given phase (system) consists of two components: (a) energy per unit mass of the phase, u^m and (b) energy per unit surface area of the phase, u^s.

The total internal energy of the given phase (system) is therefore:

$$U = u^m m + u^s A \tag{1.1}$$

where m is the mass of the phase, and A is its surface area. From equation (1.1) we get the internal energy per unit mass:

$$\frac{U}{m} = u^m + u^s \frac{A}{m} \tag{1.2}$$

where A/m is the surface area per unit mass (the specific surface area). In the case of systems with weakly developed surfaces, the quotient A/m assumes a very small value, and hence the second term in the right-hand side of equation (1.2) can be neglected. When the surface of a system is developed (e.g. as a result of high diminution) the specific surface area assumes a large value, and the surface energy (the interfacial energy) has a significant effect on the properties of the given system.

The unbalanced forces at the phase boundary (the so-called surface forces) cause changes in the number of molecules (atoms, ions) to occur on the boundary surface as compared with the corresponding numbers within the neighbouring (gaseous or liquid) phases. This change in concentration at the surface is referred to as adsorption, and may be a physical process (physisorption) taking place due to van der Waals forces, hydrogen bonding, etc., or it may be due to chemical processes and to the formation of chemical compounds (chemical adsorption or chemisorption). As a result of these processes, the balance of forces at the interface is partly or fully restored.

We should note that adsorption is to be distinguished from absorption since the latter consists of the penetration of a substance from one phase into the bulk of another by diffusion. One should clearly distinguish adsorption and absorption and use the appropriate name. Frequently, however, when adsorption and absorption processes occur simultaneously, it is difficult to ascertain which is dominant, and in such cases we use the general term sorption.

Adsorption processes are usually classified according to the kind of phases constituting the interface, and according to the type of forces acting at this surface.

Depending on the type of phases in contact, we can consider the process of adsorption in the following systems:

(1) liquid/gas, (3) solid/liquid,
(2) solid/gas, (4) liquid/liquid.

In subsequent chapter we shall successively discuss adsorption in the above systems. The classification of adsorption processes in terms of interfacial forces will be treated in a separate chapter.

Chapter **2**

Adsorption at the Liquid/Gas Interface

2.1 Surface Tension of Solutions

Cohesive forces between molecules are unbalanced at a phase boundary. In the case of the liquid/gas interface the resultant force acting on the surface molecules from the liquid side exceeds that from the gas side (Fig. 2.1). Surface molecules of the liquid are therefore drawn into the liquid phase,

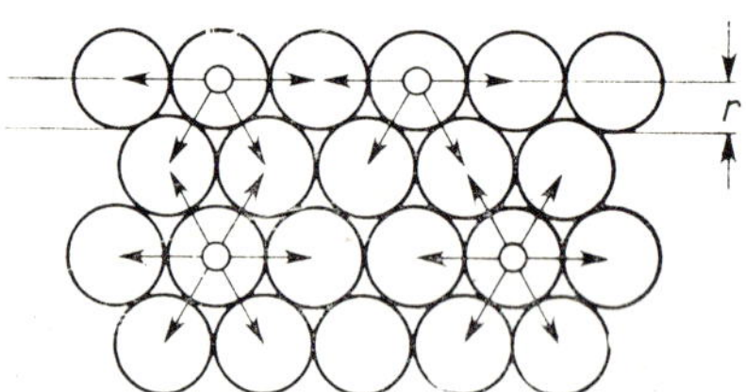

FIG. 2.1 Surface layer at the liquid/gas interface (r is the radius of intermolecular interaction).

whose surface area tends to diminish. As a consequence of interactions among liquid molecules, the force directed normal to the surface into the liquid is accompanied by another force directed tangentially to the surface and preventing its area from increasing. This latter force per unit length (the cross section of the surface is a line) is a measure of the surface tension (denoted by γ or σ).

Since the liquid tends to diminish its surface area, increase of that area requires work against the surface tension. Figure 2.2 shows a hypothetical system consisting of a liquid filling a container having a sliding cover. It is assumed that the nature of the lid is such that there is no surface tension between it and the liquid. We shift the lid to uncover a part of the liquid surface of area equal to dA. If the operation is performed quasistatically at constant temperature and pressure, the work will be equal to the increase

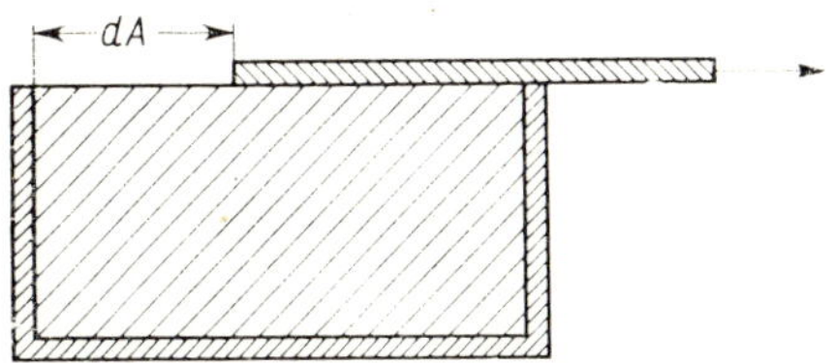

FIG. 2.2 Formation of a new surface of the liquid.

of Gibbs free energy:

$$\mathrm{d}G^{\mathrm{s}}=\gamma\,\mathrm{d}A \tag{2.1}$$

where G^{s} is the Gibbs free energy of formation of the new area of surface. In general:

$$G^{\mathrm{s}}=\gamma A \tag{2.2}$$

or

$$\gamma=\frac{G^{\mathrm{s}}}{A}\,. \tag{2.3}$$

It follows that the units of surface tension are N/m or J/m^2 (dyne/cm or erg/cm^2 in the CGS system). Surface tension represents the work necessary to form unit area of new surface or, equivalently, the increase of Gibbs free energy corresponding to the formation of unit area of surface.

In accordance with the second law of thermodynamics the Gibbs free energy tends to a minimum (under the given conditions of constant temperature and pressure). It follows from equation (2.2) that the surface free energy may change in two ways:

(i) As a result of changes in surface area A only, as in the case of pure liquids, where the surface tension depends only on the nature of the liquid.

(ii) As a result of changes in both surface area and surface tension, as in the case of solutions.

In the case of mixtures of two liquids with similar surface tensions, we observe an almost linear variation of surface tension with concentration.

In so-called regular solutions [1], the surface tension can be represented in the form derived by Prigogine and Defay [2]:

$$\gamma=\gamma_1 x_1+\gamma_2 x_2-\beta_{\mathrm{p}} x_1 x_2 \tag{2.4}$$

where γ_1 and γ_2 are the surface tensions of the pure liquids, x_1 and x_2 are their mole fractions and β_p is a semi-empirical constant.

For ideal solutions the term $\beta_p x_1 x_2$ in equation (2.4) should be equal zero over the whole concentration range. In Fig. 2.3 a typical example is presented for the chloroform/acetone system.

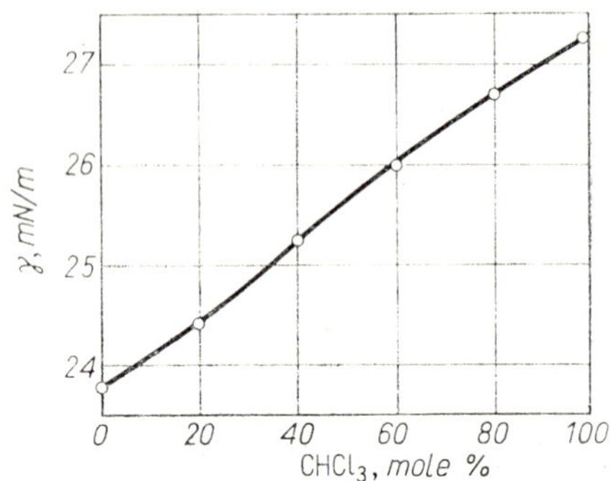

FIG. 2.3 Variation of the surface tension with composition of the chloroform–acetone solution at 291 K (after Adamson [3]).

If the surface tensions of both liquids differ significantly, the addition of a small amount of liquid of lower surface tension usually lowers the surface tension of the solution as compared with that of the solvent (Fig. 2.4). In the case of dilute aqueous solutions of organic substances which lower the surface tension, an empirical equation derived by Szyszkowski [4] is used. This equation describes the relationship between the surface tension

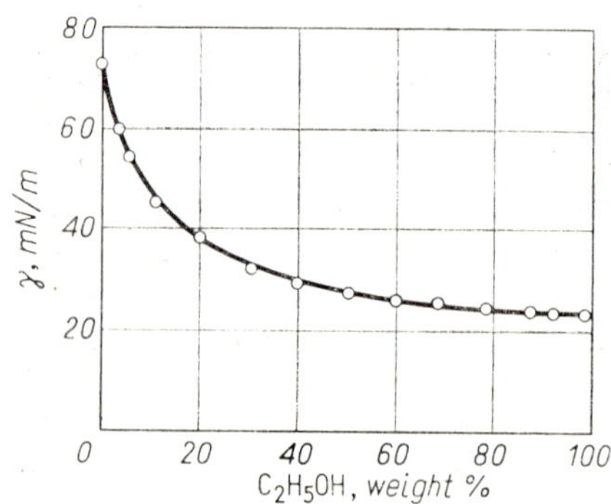

FIG. 2.4 Variation of surface tension with composition of aqueous ethanol at 298 K (after Adamson [3]).

γ_0 of the solvent, the surface tension γ of the solution, and its concentration c,

$$\frac{\gamma_0-\gamma}{\gamma_0}=B\ln\left(1+\frac{c}{A}\right) \tag{2.5}$$

or, equivalently,

$$\gamma_0-\gamma=b\ln(1+ac) \tag{2.6}$$

where B and b are constants specific for the given homologous series and $a=1/A$ is the specific capillary activity characteristic of the given compound.

Szyszkowski derived his equation in the course of studies on surface tension of aqueous solutions of fatty acids (with C_3 to C_6 molecules). This equation is only fulfilled for some simple cases, but it does deserve attention because of its theoretical interpretation (see Section 2.4).

Substances which, in small concentrations, cause a lowering of the surface tension of a solution are referred to as capillary or surface active substances. Many substances are surface active with respect to water, since their surface tension is lower than that of water, whose surface tension is high. These include such organic compounds as alcohols, fatty acids, esters, ethers, etc.

In 1884, Traube conducted an extensive study [5] of the variation with concentration of the surface tension of aqueous solutions of various surface active substances. He found that the capillary activity of members of a homologous series increases regularly with chain length, such that every additional —CH_2— group increases the capillary activity 3.2 times. In other words, for equal lowering of the surface tension in a homologous series, e.g. by $\Delta\gamma$, it is sufficient to use the successive member of the series with a concentration 3.2 times lower. This regularity, fulfilled for dilute solutions, is known as the Traube rule. In the case of equal $\Delta\gamma$ values for two members of a homologous series, we can write in accordance with the Szyszkowski equation that

$$b_n\ln\left(1+\frac{c_n}{A_n}\right)=b_{n+1}\ln\left(1+\frac{c_{n+1}}{A_{n+1}}\right) \tag{2.7}$$

where n is the number of the homologue in the series. Since $b_n=b_{n+1}$, the Traube rule gives

$$\frac{A_n}{A_{n+1}}=\frac{a_{n+1}}{a_n}=\frac{c_n}{c_{n+1}}=3.2\,. \tag{2.8}$$

Figure 2.5 presents a range of isotherms of γ-dependence on concentration for aqueous fatty acid solutions. The isotherm becomes steeper with increasing numbers of carbon atoms within the homologous series.

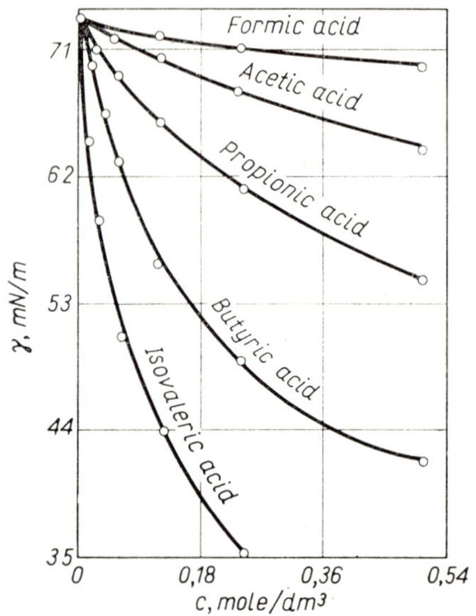

FIG. 2.5 Isotherms of surface tension variation with concentration of acid aqueous solutions for various fatty acids.

If the solvent is a liquid of low surface tension, then those substances which cause surface tension lowering in aqueous solutions may now produce an increase and thus no longer be surface active. The surface activity of a given substance is therefore a function not only of its properties but also of the solvent.

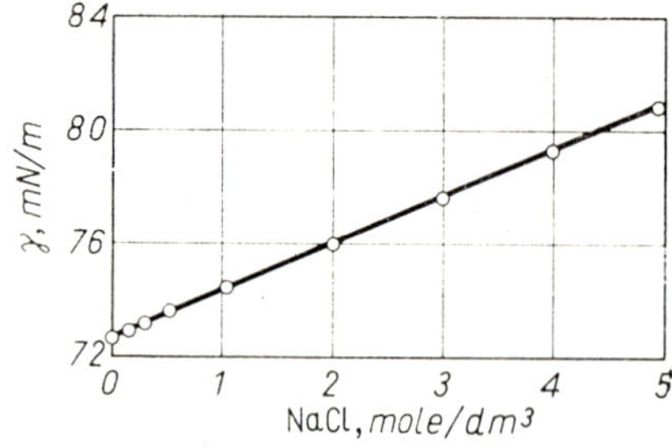

FIG. 2.6 Variation of surface tension of NaCl aqueous solution at 293 K with concentration (after Adamson [3]).

The surface tension of electrolytes is somewhat higher than that of pure water and increases with concentration (Fig. 2.6).

Colloidal electrolytes (e.g. sodium myristate) form a separate group of aqueous solutions. Here the surface tension decreases rapidly for very low concentrations of these substances in solution, and further increase in concentration has only a minor effect on the surface tension (Fig. 2.7).

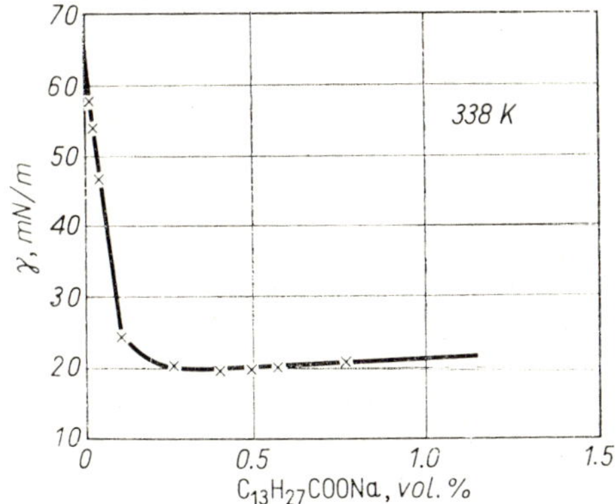

FIG. 2.7 Variation of surface tension with concentration of sodium myristate in aqueous solution (after Lottermoser and Tasch [6]).

The explanation of such a concentration dependence may be found in their capacity to form micelles (larger and more complicated particles) in more concentrated solutions. As a result the concentration of the surface active substance ceases to increase as the bulk concentration increases. Thus the surface tension becomes constant, independent of concentration.

2.2 Adsorption at the Surface of Solutions

2.2.1 Surface Excess Amount

Gibbs (1878) first considered in detail the thermodynamics of adsorption at the solution/gas interface. His methods and results are not restricted to solution surfaces, and may be regarded as the foundation of the physical chemistry of surface phenomena. In view of its importance, we shall review his work here.

Figure 2.8 shows two identical systems (I and II) consisting of phases α and β separated by the surface XY. n_i moles of substance i introduced into system I will distribute between the two phases. Denote by c_i^α and c_i^β the concentration of substance i in the α-phase and β-phase, respectively. Because of adsorption the concentration profile of substance i as a function

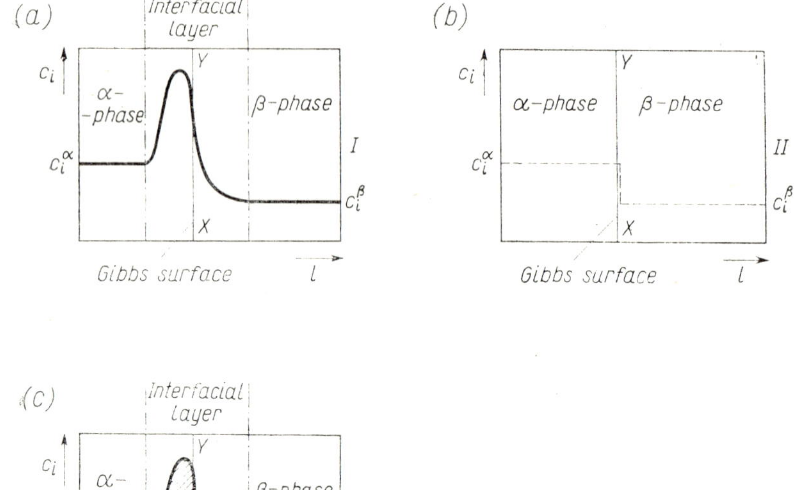

FIG. 2.8 Adsorption at the α-phase/β-phase interface (after [7]): (a) real system I where the solid line is the concentration profile of substance i as a function of the distance from the XY surface (Gibbs surface); (b) reference system II where no adsorption occurs, the dashed line is the concentration profile of substance i as a function of the distance from the XY surface; (c) comparison of systems I and II showing the surface excess amount of substance i (n_i^σ) (shaded area).

of distance from surface XY has the shape shown in the figure by the solid line (distance from the XY phase boundary varies horizontally and concentration vertically). In the phase boundary region the concentration of substance i differs from that in the bulk phases at a sufficiently large distance from the XY phase boundary. This phase boundary region is called the surface or interfacial layer or the surface phase.

Let us assume that no adsorption occurs in system II, and introduce into that reference system a number of moles of substance i such that after distribution between the α and β phases the concentrations in the bulk phases are c_i^α and c_i^β, respectively (i.e. the same as in system I). Owing to lack of adsorption the concentration profile of substance i as a function of distance from surface XY has the form shown in the figure by the dashed line. In this case the concentration of substance i in both phases remains unchanged down to the surface XY. In that system the number of moles of substance i in phase α is $n_i^\alpha = V^\alpha c_i^\alpha$ and in phase β is $n_i^\beta = V^\beta c_i^\beta$ (where V^α and V^β are the volumes of phases α and β, respectively).

If Figs. 2.8a and 2.8b are superimposed, we find (Fig. 2.8c) that in system I (real system) there is a certain excess of moles of substance i as compared with the number of moles in system II (reference system). In Fig. 2.8c this excess amount is represented by the shaded surface area and is given by:

$$n_i^\sigma = n_i - (V^\alpha c_i^\alpha + V^\beta c_i^\beta)\,. \tag{2.9}$$

The quantity n_i^σ is the surface excess amount or Gibbs adsorption [7]. If we divide n_i^σ by the surface area A of the boundary surface XY, we obtain a quantity called by Gibbs the surface concentration (measured in mole/m^2) and denoted by Γ_i^σ,

$$\Gamma_i^\sigma = n_i^\sigma / A\,. \tag{2.10}$$

The term surface concentration is, however, confusing since it seems to denote the total concentration of substance i in the surface layer, whereas in reality it represents the surface excess concentration (which can be positive or negative in the mathematical sense) of that substance with regard to its concentrations in the neighbouring bulk phases.

Similarly, we may define surface excess amounts of other substances present in the system. The total surface excess amount n^σ and total surface concentration Γ^σ are given by:

$$n^\sigma = \sum_i n_i^\sigma \tag{2.11}$$

and

$$\Gamma^\sigma = \sum_i \Gamma_i^\sigma. \tag{2.12}$$

The boundary surface XY may, in principle, be located arbitrarily in the system. Its position obviously influences the number of moles of substance i in the reference system (at given concentrations c_i^α and c_i^β in the two phases) and hence also the quantities n_i^σ and Γ_i^σ. For this reason the surface excess amounts, which are independent of the location of the XY boundary surface, are also well-defined.

NOTE. The boundary surface, XY, (the Gibbs surface), separating the phases α and β, is not a true boundary in the case of liquid/gas or liquid/liquid interfaces. There is no step transition of the two phases, α and β, one into the other, and, as shown in Figs. 2.8a and 2.8c, an interfacial (surface) layer exists. The Gibbs surface is only a hypothetical, geometric cross-section of that layer necessary to define the volumes of the neighbouring (bulk) phases and to subsequently calculate adsorption and other surface excesses [7].

2.2.2 Ways of Expressing the Surface Excess Amount

Gibbs assumed that the position of the XY surface could be chosen so that the surface excess amount of component 1 (main component, i.e. solvent, water in the case of aqueous solutions) be equal to zero. Thus the position of the Gibbs surface is such that in the reference system (system II) the number of moles of component 1 is the same as in the real system (system I). This gives the excess amount (relative adsorption) of component i with respect to component 1. Defay, Prigogine, Bellemans and Everett [8], as well as Goodrich [9], have derived an equation for this relative have adsorption [10]:

$$n_i^{\sigma(1)}=(n_i-c_i^\alpha V)-(n_1-c_1^\beta V)\frac{c_i^\alpha-c_i^\beta}{c_1^\alpha-c_1^\beta} \tag{2.13}$$

where n_1 and n_i are the numbers of moles of components 1 and i, and c_i^α, c_i^β, c_1^α, c_1^β are the concentrations of components i and 1 in the bulk phases α and β and V is the total volume of the system.

According to Goodrich [9] we can also write:

$$n_i^{\sigma(1)}=n_i-(c_i^\alpha V^{\alpha(1)}+c_i^\beta V^{\beta(1)}). \tag{2.14}$$

Equation (2.14) is obtained from equation (2.13) by transforming the right-hand side and introducing auxiliary quantities:

$$\left.\begin{aligned} V^{\alpha(1)}&=\frac{n_1-c_1^\beta V}{c_1^\alpha-c_1^\beta} \\ V^{\beta(1)}&=V-V^{\alpha(1)}=\frac{c_1^\alpha V-n_1}{c_1^\alpha-c_1^\beta} \end{aligned}\right\} \tag{2.15}$$

where $V^{\alpha(1)}$ and $V^{\beta(1)}$ are the relative volumes of phases α and β with respect to component 1. Volume $V^{\alpha(1)}$ is equal to that of phase α with fictitious properties whose concentration c_1^α is constant up to the Gibbs surface. The volume $V^{\beta(1)}$ can be defined in a similar manner.

Dividing $n_i^{\sigma(1)}$ by the Gibbs surface area A, we get the surface excess concentration of component i with respect to component 1:

$$\Gamma_i^{(1)} = n_i^{\sigma(1)}/A. \tag{2.16}$$

The Gibbs surface excess concentration is inconvenient, since it is readily understood physically only for small concentrations of strongly adsorbable substances. For that reason, Guggenheim and Adam [11] have defined a number of other surface excess amounts.

The reduced adsorption $n_i^{\sigma(n)}$ of component i is defined by the equation [7]:

$$n_i^{\sigma(n)} = n_i^\sigma - n^\sigma \left(\frac{c_i^\alpha - c_i^\beta}{c^\alpha - c^\beta} \right) \tag{2.17}$$

where:

$$c^\alpha = \sum_i c_i^\alpha, \qquad c^\beta = \sum_i c_i^\beta.$$

The quantity $\Gamma_i^{(n)}$ can be defined similarly by dividing $n_i^{\sigma(n)}$ by the interfacial surface area:

$$\Gamma_i^{(n)} = n_i^{\sigma(n)}/A. \tag{2.18}$$

The reduced adsorption is independent of the location of the Gibbs surface, and can be represented as the Gibbs adsorption n_i^σ (or Γ_i^σ) of substance i when the surface XY is localized in the system so that the total surface excess n^σ (or Γ^σ) (equations (2.11) and (2.12)) is equal to zero. This means that the Gibbs surface has been chosen so that the reference and real systems contain the same total number of moles (n) of the components. Thus:

$$\sum_i n_i^{\sigma(n)} = \sum_i \Gamma_i^{(n)} = 0. \tag{2.19}$$

The reduced adsorption can also be calculated from the equation:

$$n_i^{\sigma(n)} = n_i - (c_i^\alpha V^{\alpha(n)} + c_i^\beta V^{\beta(n)}) \tag{2.20}$$

where:

$$V^{\alpha(n)} = \frac{n - c^\beta V}{c^\alpha - c^\beta}, \qquad V^{\beta(n)} = \frac{c^\beta V - n}{c^\alpha - c^\beta}$$

and

$$n = \sum_i n_i.$$

The surface excess amounts $n_i^{\sigma(v)}$ and $\Gamma_i^{(v)}$ as well as $n_i^{\sigma(m)}$ and $\Gamma_i^{(m)}$ are defined similarly. In the case of $n_i^{\sigma(v)}$ and $\Gamma_i^{(n)}$, we compare the number of moles of substance i in the real system with that of the same substance in the reference system with the same total volume (of course at T=const). In order to calculate $n_i^{\sigma(m)}$ and $\Gamma_i^{(m)}$ we compare the number of moles of substance i in the real system with that of the same substance in the reference system with the same total mass.

The surface excess amounts given by Guggenheim and Adam can be related by the equation:

$$\sum_i P_i \Gamma_i = 0 \tag{2.21}$$

where the constants P_i depend on the definition of the surface excess amount, i.e. on the choice of Gibbs surface in Fig. 2.8. For $\Gamma_i^{(n)}$, $P_i=1$; for $\Gamma_i^{(m)}$, $P_i=M_i$ (where M_i is the molecular weight); for $\Gamma_i^{(v)}$, $P_i=V_i$ (where V_i is the molar volume).

Equation (2.21), the Guggenheim and Adam equation, is important in the description of adsorption for solutions. It expresses the obvious fact that the number of molecules of a given substance may increase in the surface layer only at the expense of an equal number of molecules of other components of the solution and vice versa.

Simple relations exist between the Gibbs surface excess, and those derived by Guggenheim and Adam. It can easily be shown [3] that in the case of a two-component solution:

$$x_1 \Gamma_2^{(1)} = \Gamma_2^{(n)} = \frac{\bar{M}}{M_1} \Gamma_2^{(m)} = \frac{\bar{V}}{V_1} \Gamma_2^{(v)} \tag{2.22}$$

where x_1 (or x_2) is the mole fraction of the given component of the solution, $\bar{M}=x_1 M_1 + x_2 M_2$ is the average molecular weight of the solution, and $\bar{V}=x_1 V_1 + x_2 V_2$ is the average molar volume of the solution.

When $x_2 \to 0$ then x_1, $\bar{M}/M_1$ and $\bar{V}/V_1$ tend to unity, and

$$\Gamma_2^{(1)} = \Gamma_2^{(n)} = \Gamma_2^{(m)} = \Gamma_2^{(v)}.$$

For sufficiently dilute solutions all the excess amounts display a linear concentration dependence:

$$\Gamma_2 = k_2 x_2 . \tag{2.23}$$

Similarly, for the case of solvents, we have, for $x_1 \to \infty$:

$$\Gamma_1^{(2)} = \Gamma_1^{(n)} = \Gamma_1^{(m)} = \Gamma_1^{(v)} = k_1 x_1 . \tag{2.24}$$

In the limiting case $x_2 \to 1$ we have for the dissolved substance:

$$\Gamma_2^{(n)} = -\Gamma_1^{(n)} = -k_1 x_1, \tag{2.25}$$

$$\Gamma_2^{(m)} = -\frac{M_1}{M_2}\Gamma_1^{(m)} = -\frac{M_1}{M_2}k_1 x_1, \tag{2.26}$$

$$\Gamma_2^{(v)} = -\frac{V_1}{V_2}\Gamma_1^{(v)} = -\frac{V_1}{V_2}k_1 x_1. \tag{2.27}$$

In this limiting case, $\Gamma_2^{(1)}$ reveals a different behaviour. From equation (2.22) and its equivalent for $\Gamma_1^{(2)}$ we have:

$$x_1 \Gamma_2^{(1)} = -x_2 \Gamma_1^{(2)} = -\Gamma_1^{(n)}. \tag{2.28}$$

When $x_2 \to 1$, then

$$x_1 \Gamma_2^{(1)} = -k_1 x_1 \tag{2.29}$$

and

$$\Gamma_2^{(1)} = -k_1 .$$

Similarly, when $x_1 \to 1$, then

$$\Gamma_1^{(2)} = -k_2 . \tag{2.30}$$

This shows that, contrary to intuition, $\Gamma_2^{(1)}$ does not tend to zero when the composition of the solution tends to the pure solute.

2.3 The Gibbs Adsorption Isotherm

2.3.1 Introduction

In order to derive a relationship between adsorption at the solution/gas interface and the solution composition we shall consider the thermodynamic equilibrium of such a system. Let us assume that in the system in Fig. 2.8 the liquid (solution) is the α-phase and the gas (vapour) is the β-phase.

If we consider a virtual (imaginary) change of components in the surface layer, we can write in accordance with the definition of Helmholtz free energy ($V^\sigma = 0$):

$$\mathrm{d}F = -S^\sigma \mathrm{d}t + \gamma \mathrm{d}A + \sum_i \mu_i \mathrm{d}n_i^\sigma . \tag{2.31}$$

The chemical potentials are not provided with superscripts since at thermodynamic equilibrium the chemical potential of every component of a system is the same in every phase, i.e. $\mu_i^\alpha=\mu_i^\beta=\mu_i^\sigma=\mu_i$. At constant temperature and pressure equation (2.31) assumes the form:

$$dF^\sigma=\gamma dA+\sum_i \mu_i dn_i^\sigma. \tag{2.32}$$

We integrate equation (2.32) between the limits of zero and the true value of the surface area A. This is equivalent to increasing the surface layer from zero to A, at constant composition. After integrating we get:

$$F^\sigma=\gamma A+\sum_i \mu_i n_i^\sigma. \tag{2.33}$$

On differentiating equation (2.33) we obtain:

$$dF^\sigma=\gamma dA+A d\gamma+\sum_i \mu_i dn_i^\sigma+\sum_i n_i^\sigma d\mu_i. \tag{2.34}$$

Substracting equation (2.32) from equation (2.34) gives:

$$A d\gamma+\sum_i n_i^\sigma d\mu_i=0 \tag{2.35}$$

or

$$d\gamma=-\sum_i \frac{n_i^\sigma}{A} d\mu_i \tag{2.36}$$

or, in terms of the Gibbs surface concentration, defined by equation (2.9):

$$d\gamma=-\sum_i \Gamma_i^\sigma d\mu_i. \tag{2.37}$$

This latter equation, known as the Gibbs adsorption equation, relates variations of surface tension to the Gibbs surface concentrations and variations of the chemical potentials of the solution components.

For a two-component system, equation (2.37) takes the form:

$$d\gamma=-\Gamma_1^\sigma d\mu_i-\Gamma_2^\sigma d\mu_2. \tag{2.38}$$

When the phases in contact are: gas (vapour) and solution 1+2, and assuming that component 1 (solvent) is present in large excess in the solution, in accordance with the Gibbs assumption we have:

$$\Gamma_1^\sigma=0.$$

Then:

$$d\gamma=-\Gamma_2^{(1)} d\mu_2. \tag{2.39}$$

Since

$$\mu_2=\mu_2^\ominus+RT\ln a_2 \tag{2.40}$$

then

$$d\gamma = -\Gamma_2^{(1)} RT \, d \ln a_2 \tag{2.41}$$

or

$$\Gamma_2^{(1)} = -\frac{1}{RT}\left(\frac{\partial \gamma}{\partial \ln a_2}\right)_A \tag{2.42}$$

or, equivalently,

$$\Gamma_2^{(1)} = -\frac{a_2}{RT}\left(\frac{\partial \gamma}{\partial a_2}\right)_A. \tag{2.43}$$

Equation (2.42) or (2.43) is known as the Gibbs adsorption isotherm, relating the surface excess concentration $\Gamma_2^{(1)}$ to the variation of γ with solute activity a_2. For dilute solutions, we can replace the activity a_2 by concentration x_2, when:

$$\Gamma_2^{(1)} = -\frac{x_2}{RT}\left(\frac{\partial \gamma}{\partial x_2}\right)_A = -\frac{1}{RT}\left(\frac{\partial \gamma}{\partial \ln x_2}\right)_A. \tag{2.44}$$

If the solute concentration is given in moles per litre (c_2), we can write:

$$\Gamma_2^{(1)} = -\frac{c_2}{RT}\left(\frac{\partial \gamma}{\partial c_2}\right)_A = -\frac{1}{RT}\left(\frac{\partial \gamma}{\partial \ln c_2}\right)_A. \tag{2.45}$$

If component 1 constitutes the liquid phase (e.g. water) and component 2, insoluble in the liquid phase (e.g. saturated hydrocarbon), constitutes the gas phase, then the chemical potential of the liquid phase remains constant and $d\mu_1 = 0$. In this case equation (2.38) can be written in the form:

$$d\gamma = -\Gamma_2^{\sigma} \, d\mu_2 . \tag{2.46}$$

If component 2 behaves as an ideal gas whose pressure is p_2, then

$$d\mu_2 = RT \, d\ln p_2 \tag{2.47}$$

and hence

$$d\gamma = -\Gamma_2^{\sigma} RT \, d\ln p_2 \tag{2.48}$$

or

$$\Gamma_2^{\sigma} = -\frac{1}{RT}\left(\frac{\partial \gamma}{\partial \ln p_2}\right)_A = -\frac{p_2}{RT}\left(\frac{\partial \gamma}{\partial p_2}\right)_A. \tag{2.49}$$

We have thus obtained the Gibbs adsorption isotherm for gas (vapour) adsorption on the liquid surface.

From the Gibbs adsorption isotherm we have the following conclusions:

(i) If the solute lowers the surface tension of the solution as its concentration increases (surface active compound) then

$$\left(\frac{\partial \gamma}{\partial c_2}\right)_A < 0 \text{ and } \Gamma_2^{(1)} > 0$$

which means that the surface active compound accumulates in the surface phase (positive adsorption).

(ii) If increasing concentration of the solute increases the surface tension of the solution, then

$$\left(\frac{\partial \gamma}{\partial c_2}\right)_A > 0 \text{ and } \Gamma_2^{(1)} < 0$$

which means that the solute avoids the surface phase (negative adsorption).

(iii) If the solute has no effect on the surface tension of the solution, then

$$\left(\frac{\partial \gamma}{\partial c_2}\right)_A = 0 \text{ and } \Gamma_2^{(1)} = 0$$

when the concentration in the surface phase is the same as in the remaining volume of the solution and no adsorption is observed.

2.3.2 Calculation of Adsorption Using the Gibbs Isotherm

From the behaviour of the function $\gamma = F(c_2)$ we can calculate and plot the function $\Gamma_2^{(1)} = f(c_2)$ based on the Gibbs adsorption isotherm. These calculations can be performed in different ways. Most frequently the function $\gamma = F(c_2)$ is differentiated graphically when the values of the derivatives $(\partial\gamma/\partial c_2)_A$ are obtained. These differentiations can be performed in two ways.

(i) We draw tangents to the $\gamma = F(c_2)$ curve which intersect the axis of abscissae at angles $\alpha_1, \alpha_2, \ldots,$ etc. (Fig. 2.9). The tangents of these angles are the respective values of the derivative $(\partial\gamma/\partial c_2)_A$ for various concentrations of component 2 in the solution. Multiplying the resultant $(\partial\gamma/\partial c_2)_A$ values by the corresponding expressions $-c_2/RT$ we get the desired surface excess concentrations $\Gamma_2^{(1)}$.

(ii) We draw tangents to the $\gamma = F(c_2)$ curve which intersect with the axis of ordinates (Fig. 2.10). Through the points at which tangents have been drawn, we draw straight lines parallel to the axis of abscissae to intersect the axis of ordinates. The tangents and the corresponding lines parallel

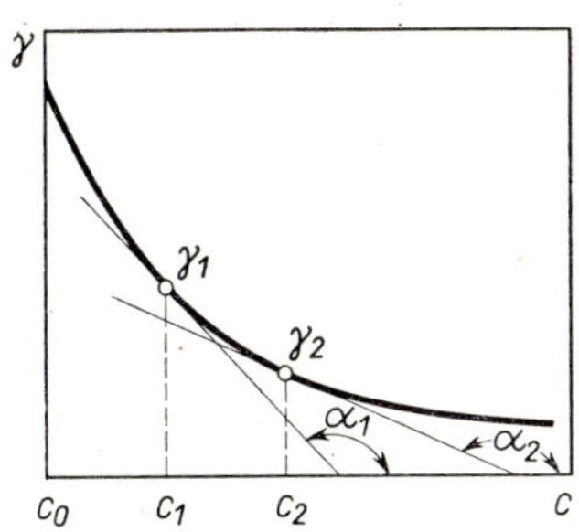

FIG. 2.9 Graphical method of determining the Gibbs adsorption isotherm $\Gamma=f(c)$ from the $\gamma=F(c)$ isotherm (determination of the tangents of angles at which the tangents to the curve intersect the abscissa).

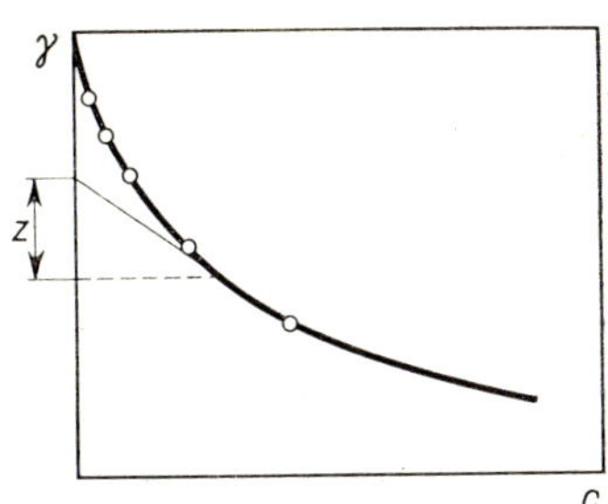

FIG. 2.10 Graphical method of determining the Gibbs adsorption isotherm $\Gamma=f(c)$ from $\gamma=F(c)$ isotherm (determination of section z on the ordinate).

to the abscissa give sections z on the axis of ordinates. From elementary geometry it follows that:

$$-\frac{z}{c_2}=\left(\frac{\partial\gamma}{\partial c_2}\right)_A$$

or

$$z=-c_2\left(\frac{\partial\gamma}{\partial c_2}\right)_A. \tag{2.50}$$

Substituting this value into equation (2.45) we get:

$$\Gamma_2^{(1)}=\frac{z}{RT}. \tag{2.51}$$

In this way we can calculate the surface excess concentration $\Gamma_2^{(1)}$ corresponding to those concentrations for which tangents have been drawn to the $\gamma=F(c_2)$ curve.

The excess concentrations $\Gamma_2^{(1)}$ can also be calculated if we assume that:

$$\left(\frac{\partial\gamma}{\partial c_2}\right)_A=\left(\frac{\Delta\gamma}{\Delta c_2}\right)_A.$$

This procedure is, however, less accurate than that based on graphical differentiation.

NOTE. If c_2 is expressed in mole/dm^3, surface tension γ in J/m^2, and if we assume that $R=8.314\ J\cdot K^{-1}\cdot\text{mole}^{-1}$, then $\Gamma_2^{(1)}$ will be obtained in mole/m^2.

2.3.3 Checking the Gibbs Equation Experimentally

An experimental check of the Gibbs equation is not easy, particularly because of the difficulties in determining experimentally the surface excess concentration of a solution.

McBain *et al.* [12], in the years 1932–36, used a microtome whose blade, moving at a speed of over 11 m/s, cut the surface layer 0.05 to 0.1 mm thick, and this layer was subsequently analysed. Some results obtained by this method are given in Table 2.1. On the whole, the experimental results of the determination of the surface excess concentration were in agreement, within experimental error, with the values calculated from the Gibbs equation on the basis of surface tension variation due to concentration changes.

Table 2.1 Values of surface excess concentration calculated and found experimentally (after McBain *et al.* [12])

Solute	Number of experiments	c_2 g/1000 g H_2O	$\Gamma_2^{(1)}$, (g/cm^2) × 10^4 exp.	theor.
p-Toluidine	11	1.85×10^{-2}	6.1	5.2
Phenol	29	1.63×10^{-2}	4.6	4.9
Caproic acid	14	2.59×10^{-2}	5.1	6.5
Phenol	18	2.18×10^{-2}	4.1	4.8
Sodium chloride	—	2.0 mole/dm^3	−0.43	−0.37

Salley, Dixon *et al.* [13, 14] developed a method of measuring Γ^σ by making use of labelled atoms. They labelled the tested substance with an isotope emitting weak β-radiation, e.g. ^{14}C or ^{35}S, and subsequently placed a radiation detector adjacent to the solution surface. Since the range of β-radiation emitted from a weak source is small, the measured radioactivity reflected the surface region together with a thin layer of the bulk phase. The values of surface excess concentration obtained by this method are in approximate agreement with those calculated on the basis of the Gibbs adsorption isotherm.

2.3.4 Final Remarks

Figure 2.11 presents the dependence of surface excess concentration $\Gamma_2^{(1)}$ of ethanol on its concentration in aqueous solution. The quantity $\Gamma_2^{(1)}$ was calculated from the Gibbs adsorption isotherm and based on the experimentally determined plot of the $\gamma = f(x_2)$ function where x_2 is the ethanol mole fraction. In this figure the dependence on the solution concentration of other surface excess concentrations of ethanol are also shown. The surface excess concentrations were calculated using relation (2.22).

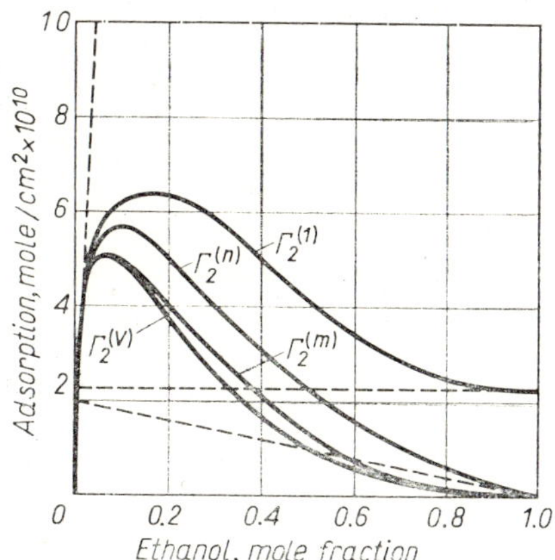

FIG. 2.11 Ethanol excess adsorption isotherms at the aqueous solution/vapour interface (after Adamson [3]).

We have, in Sections 2.2 and 2.3, discussed the problems of surface excess concentration and the Gibbs adsorption isotherm in considerable detail since they are often confused and in consequence present a source of many mistakes and errors in the interpretation of adsorption phenomena.

2.4 The Structure of the Adsorption Surface Layer

If the surface tension of a solution decreases with increasing solute concentration, then, in accordance with the Gibbs adsorption isotherm, the solute is adsorbed at the interface. Such substances, called surface active, have molecules of asymmetric structure consisting of one or more polar groups, (e.g. —OH, —COOH, —NH_2, —SO_3H) which are easily solvated, and a non-polar part (e.g. hydrocarbon chain).

Let us consider the adsorption of surface active compounds and the structure of surface layers in some detail for aqueous solutions. The interaction between molecules of water in solution is greater than that between water molecules and solute molecules (in general). As a result, the solute molecules will be pushed from the bulk of the solution to the surface, with surface excess concentration $\Gamma_2^\sigma > 0$. Accumulation of solute molecules at the surface has the effect of decreasing intermolecular interactions in the surface layer and the surface tension of the solution decreases with increasing concentration c_2 (or activity a_2).

As the size of the non-polar part of the solute molecule increases (e.g. with increasing length of carbon chain) the surface tension is already significantly lowered at small concentrations. Rebinder [15] has called the quantity $-(\partial\gamma/\partial c_2)_{c_2\to 0}$ the surface activity of the given substance.

To analyse the structure of the surface adsorption layers in more detail let us return to the Szyszkowski equation,

$$\frac{\gamma_0-\gamma}{\gamma_0}=B\ln\left(\frac{c_2}{A}+1\right). \tag{2.52}$$

By differentiating this equation with respect to c_2 we get:

$$-\frac{d\gamma}{dc_2}=\frac{B\gamma_0}{c_2+A}. \tag{2.53}$$

Substitution of equation (2.53) into the Gibbs adsorption isotherm (equation (2.45)) gives:

$$\Gamma_2^{(1)}=\frac{B\gamma_0}{RT}\,\frac{c_2}{c_2+A}. \tag{2.54}$$

For strongly adsorbed substances $\Gamma_2^{(1)}$ is very large even at very low concentrations c_2 so it eventually corresponds to the total concentration of component 2 in the surface layer (per 1 cm^2). With this simplification we can write Γ_2, deleting the superscript. Then equation (2.54) takes the form:

$$\Gamma_2=\Gamma_{2\infty}\,\frac{c_2}{c_2+A} \tag{2.55}$$

where $\Gamma_{2\infty}$ corresponds to complete coverage of the liquid/gas interface with molecules of the adsorbed component. The experimental constant B can be given an intepretation which is in qualitative agreement with experiment. It follows from equations (2.54) and (2.55) that:

$$B=\frac{\Gamma_{2\infty}RT}{\gamma_0}. \tag{2.56}$$

The experimentally-determined values of B for the members of a homologous series are very similar. This seems justified, since by equation (2.56) the value of B depends on the surface area $\omega_m=1/\Gamma_{2\infty}N$ (where N is the

Avogadro number) per molecule of adsorbed substance. The value of ω_m, calculated in terms of $RT/B\gamma_0 N$, range, for different molecules, from 0.2 to 0.4 nm^2 (i.e. 20–40 Å^2). This corresponds in general to vertical orientation of the molecules with respect to the surface in a completely filled adsorption layer. This is likely since the molecules of surface active substances are asymmetric, and include a highly lyophilic part, e.g. hydrophilic groups such as —COOH or –OH.

Taubman [16] has collected experimental data on the values of ω_m in aqueous solutions of surface active compounds with small and medium molecules (see Table 2.2). It then becomes obvious why the values of the B constant in the Szyszkowski equation are so similar even for very distant members of a homologous series.

Table 2.2 SURFACE AREAS ω_m (IN nm^2) PER MOLECULE (AFTER TAUBMAN [16])

	Acids	Alcohols	Amines
Aliphatic	0.305	0.285	0.300
Aromatic	0.300	0.285	0.280

On the other hand, the Traube rule establishes a relationship between the surface activity and chain length. Langmuir has given the following interpretation of that rule [17]: the work required to transfer 1 mole of a substance from the bulk phase (solution) to the surface layer is in thermodynamic terms:

$$W^s = RT \ln \frac{c^s}{c} \tag{2.57}$$

where c^s is the concentration in the surface layer. For two members of a homologous series with n and $n+1$ carbon atoms in the chain we get:

$$W^s_{n+1} - W^s_n = RT \ln \frac{c^s_{n+1}\, c_n}{c^s_n\, c_{n+1}}. \tag{2.58}$$

According to the Traube rule the surface tensions of solutions are equal if the ratio $c_n/c_{n+1} = 3.2$; then of course $c^s_n = c^s_{n+1}$. In such a case for $T = 239$ K we can write:

$$W^s_{n+1} - W^s_n = RT \ln 3.2 = 2930 \text{ J/mole (or 700 cal/mole)}.$$

The value of 2930 J/mole can be regarded as the work necessary to transfer one CH_2 group (in molar terms) from the bulk of the solution to the surface.

The value per CH_2 group being independent of chain length, we can conclude that probably all the CH_2 groups have a similar position on the surface, i.e. they lie flat on the surface of the liquid.

Traube's rule is only true for highly dilute solutions when the surface tension decreases almost linearly with increasing solute concentration.

In summary, we can state that, in the case of aqueous solutions of low concentration, the molecules of the surface active compound are initially oriented horizontally on the surface of the solution, since not only the polar groups, like —OH or —COOH, but also the non-polar ones are attracted more by the liquid phase than by the gas phase (Fig. 2.12a). With increasing solution concentration the molecules in the surface layer undergo preorientation, forming a more tightly packed layer in which their polar groups are immersed in the solution and the non-polar ones are directed vertically or are inclined with respect to the surface (Fig. 2.12b). Finally a molecular brush is formed which covers the whole surface of the solution (Fig. 2.12c).

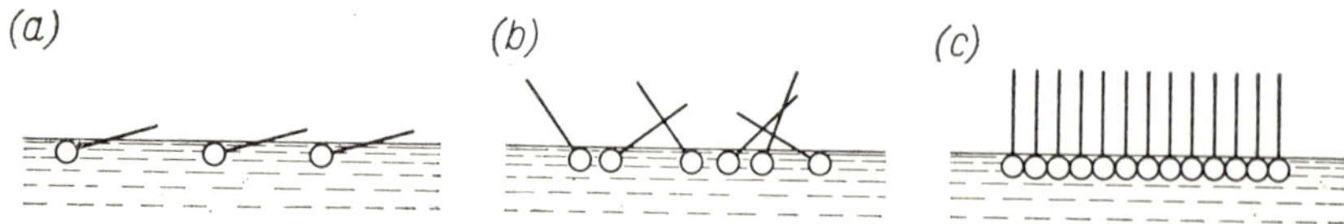

FIG. 2.12 Diagram of the structure of surface layers at the solution/vapour interface (a) for highly dilute solutions, (b) for solutions of higher concentration, (c) for complete coverage of the surface layer.

If inorganic salts are dissolved in water, then the ion–water interactions are stronger than those between water molecules. This is the reason why the ions are distributed in the bulk of the aqueous solution. The concentration of water in the surface layer is affected only slightly, however, since the ions penetrating into that layer are strongly hydrated (strong interaction between the ions and water molecules). The increased strength of intermolecular interactions in the surface layer causes a slight increase of the surface tension γ of the solution. The adsorption (negative) of ions is also quite small.

2.5 Effect of Temperature on the Surface Tension of Solutions

The surface tension of pure liquids is known to decrease linearly with temperature in accordance with the Eötvös equation [18]. However, the temperature dependence of surface tension is completely different in the case of solutions of surface active compounds:

(i) surface tension decreases linearly with temperature,

(ii) the surface excess amount n^{σ} decreases with increasing temperature, the surface layer becomes deficient in the surface active component and as a result γ increases.

References

[1] Guggenheim, E. A., *Thermodynamics*, North Holland Publishing Co., Amsterdam 1949; Longuet-Higgins, H. C., *Proc. Roy. Soc.*, A **205**, 247 (1951); Prigogine, I., *The Molecular Structure of Solutions*, North Holland Publishing Co., Amsterdam 1957.

[2] Prigogine, I. and Defay, R., *J. Chem. Phys.*, **46**, 367 (1940).

[3] Adamson, A. W., *Physical Chemistry of Surfaces*, Interscience Publishers, Ltd., London 1960.

[4] Szyszkowski, B., *Z. physik. Chem.*, **64**, 385 (1908).

[5] Traube, J., *Ann.*, **265**, 27 (1891).

[6] Lottermoser, A. and Tesch, W., *Koll.-Beih.*, **34**, Heft 5–9 (1932).

[7] Manual of Definitions, Terminology and Symbols in Colloid and Surface Chemistry, *Pure and Appl. Chem.*, **31**, 577 (1972).

[8] Defay, R., Prigogine, I., Bellemans, A. and Everett, D. H., *Surface Tension and Adsorption*, Longmans, London 1966.

[9] Goodrich, F. C., *Trans. Faraday Soc.*, **64**, 3403 (1968); *Surface and Colloid Science*, E. Matijevič (Ed.), Vol. I, Wiley–Interscience, New York 1969.

[10] Wagner, C., *Nach. Akad. Wiss., Göttingen, II Math.-Physik. Klasse*, No. 3, 37 (1973).

[11] Guggenheim, E. A. and Adam, N. K., *Proc. Roy. Soc.*, A **139**, 218 (1933).

[12] McBain, J. W. and Humpreys, R. C., *Proc. Roy. Soc.*, A **154**, 608 (1936); McBain, J. W. and Swain, R. C., *J. Phys. Chem.*, **36**, 200 (1932).

[13] Salley, D. J., Weith, A. J., Jr., Argyle, A. A. and Dixon, J. K., *Proc. Roy. Soc.*, A **203**, 42 (1950).

[14] Dixon, J. K., Judson, S. M. and Salley, D. J., *Monomolecular Layers*, Publication of the American Association for the Advancement of Science, Washington 1954, p. 63.

[15] Rebinder, P. A., *Physik. Zeitschr.*, **27**, 825 (1926); *Z. physik. Chem.*, **120**, 163 (1927).

[16] Taubman, A. B., *Dokl. Akad. Nauk. SSSR*, **71**, 343 (1950).

[17] Langmuir, J., *J. Am. Chem. Soc.*, **39**, 1848 (1917).

[18] Eötvös, R., *Wied. Ann.*, **27**, 456 (1886).

Chapter 3

Adsorption at the Solid/Gas Interface

3.1 Introduction and Description

Adsorption at the solid/gas interface is too vast a problem to permit a comprehensive treatment in a single chapter of this book [1, 2]. We shall, therefore, consider here only the basic models and theories describing this phenomenon.

Gases and vapours are generally adsorbed by solids with a highly developed surface (adsorbents). The surface area per unit mass (usually 1 g) of the adsorbent actually taking part in the adsorption process is referred to as the specific surface area, often denoted by s.

Adsorption at the solution/gas interface may be described as the accumulation of the given substance i in the surface layer (phase). In such system, there is no actual interface, so the definition of surface excess concentration requires the assumption of a hypothetical dividing surface (Gibbs surface).

In the case of adsorption at the solid/gas interface, the Gibbs surface is equivalent to the adsorbent surface. In this case we can also consider an interfacial layer consisting of two regions (Fig. 3.1): the part of the gas phase residing in the force field of the adsorbent surface—adsorption space—and the surface layer of the solid adsorbent. The surface excess amount of the adsorbed gas (Gibbs adsorption) n_i^σ is the excess number of moles of that substance present in the system over that number present in a reference system where adsorption does not occur at the same equilibrium gas pressure, i.e.

$$n_i^\sigma = \underbrace{\int (c_i^s - c_i^g)\,\mathrm{d}V}_{\text{adsorption space}} + \underbrace{\int c_i^s\,\mathrm{d}V}_{\text{surface layer of adsorbent}} \tag{3.1}$$

where c_i^s is the local concentration of substance i in a volume element dV of the interfacial layer, c_i^g is the concentration of that substance in the bulk of the gas phase. The second term of equation (3.1) is usually assumed to be

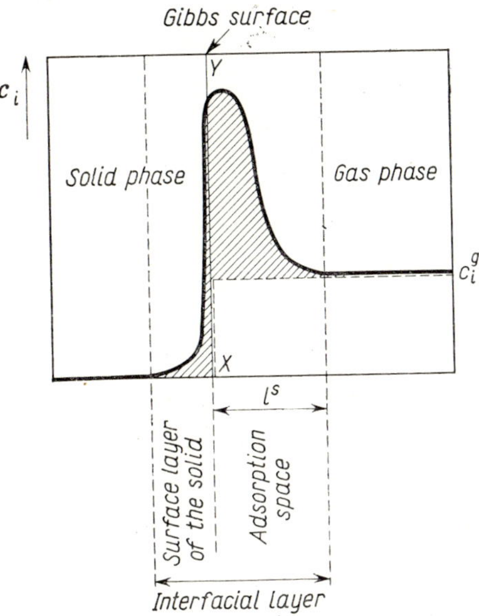

FIG. 3.1 Adsorption at the solid/gas interface. Concentration profile of substance i as a function of the distance from the solid surface: full line – in the real system, broken line – in the reference system. The surface excess amount n_i^σ is given by the shaded area.

zero, and hence:

$$n_i^\sigma = \int (c_i^s - c_i^g)\, dV. \qquad (3.2)$$

The gas concentration profile as a function of the distance normal to the adsorbent surface is shown in Fig. 3.1, where the shaded area represents n_i^σ.

For a multicomponent gas mixture the total surface excess amount of adsorbed substances is given by:

$$n^\sigma = \sum_i n_i^\sigma. \qquad (3.3)$$

If n_i^σ is the surface excess of substance i per 1 g of adsorbent whose specific surface area is s, then

$$\Gamma_i^\sigma = \frac{n_i^\sigma}{s}. \qquad (3.4)$$

The quantity n_i^σ can also be defined by:

$$n_i^\sigma = n_i - c_i^g V^g \tag{3.5}$$

where n_i is the total number of moles of substance i present in the system, c_i^g is its concentration in the gas phase, and V^g is the volume of gas in equilibrium with the adsorbent.

One can also determine the total amount of substance i in the surface layer with reference to 1 g of adsorbent n_i^s defined as:

$$n_i^s = \int_{V^s} c_i^s \, dV \tag{3.6}$$

where $V^s = l^s s$ is the volume of the interfacial layer (here, the volume of the adsorbed layer), l^s is the thickness of the adsorbed layer, and s is the specific surface area of the adsorbent. The quantity n_i^s can be conveniently defined as:

$$n_i^s = n_i^\sigma + c_i^g V^{s,g} \tag{3.7}$$

where $V^{s,g}$ is the volume of the adsorption layer (see Fig. 3.1).

When adsorption of substance i is appreciable and its equilibrium pressure is sufficiently small, the second term of the right-hand side of equation (3.7) is negligibly small, and we get

$$n_i^s \approx n_i^\sigma . \tag{3.8}$$

This approximation is verified in measurements of adsorption of gases and vapours under normal (low) pressures.

The symbol n_i^s (in moles per gram of adsorbent) has only recently been used to denote the amount of adsorption. In studies of adsorption at the solid/gas (vapour) interface various symbols have been used to denote the amount of adsorption. Most frequently we encounter in the literature the symbol a which is used to denote the amount of adsorbed substance (in mole/g), also the symbol v is used to denote the volume of the gas adsorbed (in cm^3/g). When discussing the basic models and theories of gas adsorption we shall use the most common notation, and thus assume that

$$n_i^s = a .$$

3.2 Physical and Chemical Adsorption

Physical adsorption (physisorption) involves intermolecular forces (van der Waals forces, hydrogen bonds, etc.), whereas chemical adsorption (chemisorption) involves valency forces as a result of the sharing of electrons by the solid (adsorbent) and the adsorbed substance (adsorbate).

The latter process is connected with the formation of a chemical compound involving the adsorbent and the primary layer of the substance adsorbed.

The distinction between physisorption and chemisorption of gases on solid adsorbents does not usually present great difficulties. Both kinds of adsorption are distinguished by:

1. Heat of adsorption – small in the case of physisorption, large (of the same order as the heat of the relevant chemical reaction) in the case of chemisorption.

2. Reversibility – the adsorbed substance can be relatively easily removed from the surface when physisorption is involved; the removal of a chemically adsorbed layer is very difficult and requires drastic measures.

3. Thickness of adsorbed layer – in the case of physisorption, under suitable conditions of temperature and pressure, adsorbed layers are formed having thicknesses of several diameters of the adsorbate molecule; in chemisorption only monolayers are formed.

We shall now concentrate on the phenomenological description of gas and vapour adsorption.

3.3 Adsorption Equilibrium

Adsorption equilibrium is established after considerable contact of the gas with the adsorbed surface, and can be represented in the general form:

$$f(a, p, T)=0 \tag{3.9}$$

where a is the quantity of substance adsorbed in the surface layer per 1 g of the adsorbent (moles per gram), p is the equilibrium pressure of the gas in the bulk phase (the equilibrium pressure), and T is the temperature.

Equation (3.9) can also be written in the form

$$a=f(p, T). \tag{3.10}$$

Since this equation depends upon three parameters, we can consider adsorption equilibrium in three ways:

(i) If T=const, the equilibrium can be represented by the adsorption isotherm:

$$a=f(p)_T. \tag{3.11}$$

(ii) If p=const, then the equilibrium is described by the adsorption isobar:

$$a=f(T)_p. \tag{3.12}$$

(iii) If a=const, then the equilibrium is described by the adsorption isostere:

$$p=f(T)_a. \tag{3.13}$$

Isobars and isosteres for the adsorption of ammonia on charcoal are presented in Fig. 3.2. The adsorption isotherm, however, is most frequently used in studies of adsorption equilibrium.

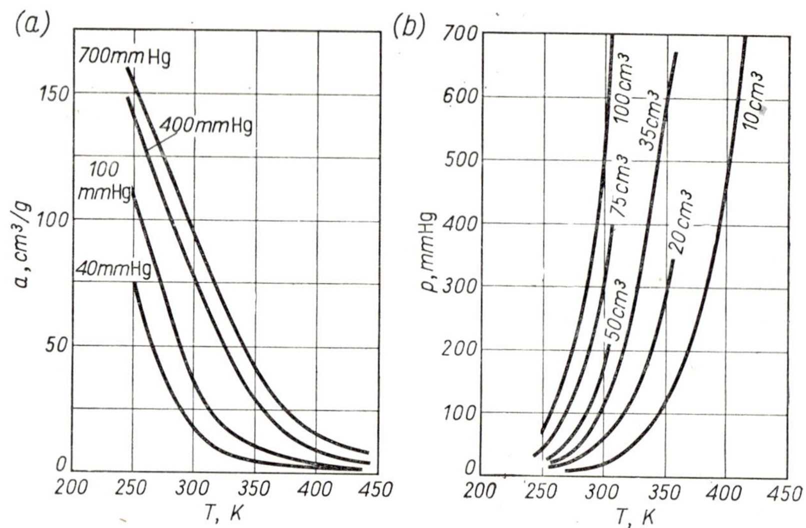

FIG. 3.2 Isobars (a) and isosteres (b) for adsorption of ammonia on charcoal (after Brunauer [1]).

3.4 Gas Adsorption on a Homogeneous Adsorbent Surface

3.4.1 Henry's Equation

Adsorption equilibrium can be represented as follows:

molecule in the gas phase ⇄ molecule on the adsorbent (adsorption complex)

In the case of a homogeneous surface, the concentration in the surface layer is constant over the whole adsorbent surface, so, for adsorption equilibrium, we have:

$$\frac{c^s f^s}{cf}=K \tag{3.14}$$

or

$$c^{s}=K\frac{f}{f^{s}}c \qquad (3.15)$$

where c^s is the concentration of the adsorbate in the surface layer (surface concentration), c is the concentration of the adsorbate in the gas phase, f^s and f are the activity coefficients in the respective phases, and K is the equilibrium constant (a function of temperature only).

Equation (3.15), relating c^s to c at constant temperature, is the adsorption isotherm equation. Since f^s and f are concentration dependent, the adsorption isotherm is nonlinear.

At low concentration in the gas phase (gas pressures up to 100 $kN \cdot m^{-2}$) we have $f \approx 1$. For small values of c^s we can also assume that $f^s \approx 1$. Since, for ideal gases, $c=p/RT$, we can write:

$$c^{s}=\frac{K}{RT}p. \qquad (3.16)$$

The total amount a of adsorbate in volume V^s of the adsorbed film per 1 g of adsorbent is calculated from the equation:

$$a=V^{s}c^{s}. \qquad (3.17)$$

From equations (3.16) and (3.17) we get:

$$a=K_{a,p}\,p \qquad (3.18)$$

where $K_{a,p}=V^sK/RT$, V^s and K represent constants under the given adsorption conditions and at constant temperature. Adsorption can also be expressed as the quantity α^s of the adsorbate per unit surface area of the adsorbent:

$$\alpha^{s}=K_{\alpha^{s},p}\,p \qquad (3.19)$$

where $K_{\alpha^{s},p}=K_{a,p}/s$.

At low pressure, a and α^s are proportional to concentration or pressure of the gas in the gas phase. This is a relation analogous to Henry's equation describing the solubility of gases in liquids.

Equations (3.16) and (3.19) illustrate different forms of the simplest adsorption isotherm equation, known as Henry's adsorption isotherm, and the corresponding constant is called Henry's constant.

At room temperature, when the gas pressure does not exceed atmospheric pressure, the adsorption isotherms for argon [3, 4], nitrogen [5, 6] and oxygen [7, 8] on activated carbon, silica gel and earths are found experimentally to be linear.

Instead of surface concentration c^s or total amount of substance adsorbed a, it is often more convenient to use the quantity θ, the surface coverage of the adsorbent:

$$\theta = \frac{c^s}{c^s_m} = \frac{a}{a_m} = \frac{\alpha^s}{\alpha^s_m} \tag{3.20}$$

where c^s_m, a_m and α^s_m are quantities corresponding to the complete coverage of the adsorbent surface by a monomolecular film of a given adsorbate.

From relations (3.16), (3.18) and (3.20) it follows that Henry's adsorption isotherm can be expressed in terms of surface coverage:

$$\theta = \frac{K}{c^s_m RT} p = \frac{K_{a,p}}{a_m} p = \frac{K_{\alpha^s,p}}{\alpha^s_m} p. \tag{3.21}$$

This means that the coverage of the adsorbent surface in Henry's zone is proportional to the pressure of the substance in the gas phase.

3.4.2 Freundlich's Adsorption Isotherm

Boedecker [9] proposed in 1895 an empirical equation for the adsorption isotherm in the form:

$$a = kp^{1/n} \tag{3.22}$$

where k and n are constants. This equation is known as Freundlich's adsorption isotherm equation, because Freundlich [10] assigned great importance to it and popularized its use. This equation, although simple and convenient, does not always accurately reproduce experimental data over a wide range of pressures. It has been widely used as an empirical equation for qualitative purposes, and seemed to have no particular theoretical foundation. A different form of this equation, proposed by Halsey [11], has proved useful in theoretical considerations of the adsorption process. Some years ago Appel [12] has derived Freundlich's adsorption isotherm using standard methods of statistical mechanics.

3.4.3 Langmuir's Theory and Adsorption Isotherm

In the years 1916–1918 Langmuir derived the adsorption theory [13] in its modern form.

On the adsorbent surface, there are a definite number of active sites (proportional to the surface area), at each of which only one molecule may be adsorbed. The bonding to the adsorbent can be either chemical or physical, but must be sufficiently strong to prevent displacement of the adsorbed molecules along the surface. We thus have localized adsorption as distinct from non-localized adsorption, where the adsorbate molecules can move freely along the adsorbent surface. We neglect interactions between the adsorbate molecules in the adsorbed film. On the surface of the adsorbent a monomolecular adsorption layer is thus formed (monolayer adsorption).

The adsorption equilibrium can then be represented as:

gas molecule+active site on the adsorbent surface⇄
localized adsorption complex.

The equilibrium constant is given by:

$$k=\frac{\alpha^{s}}{p\alpha_0^{s}}=\frac{\theta}{p\theta_0} \tag{3.23}$$

where α_0^s is the concentration of the free active sites on the adsorbent surface, $\theta_0=\alpha_0^s/\alpha_m^s$ is the fraction of the surface covered with free active sites, and α^s is the concentration of the occupied active sites on the surface, i.e. the surface concentration of the adsorbate. Of course

$$\alpha^{s}+\alpha_0^{s}=\alpha_m^{s} \tag{3.24}$$

or

$$\theta+\theta_0=1. \tag{3.25}$$

Substituting into equation (3.23) α_0^s from equation (3.24) or θ_0 from equation (3.25) we can write:

$$k=\frac{\alpha^{s}}{p(\alpha_m^{s}-\alpha^{s})}=\frac{\theta}{p(1-\theta)} \tag{3.26}$$

and after rearrangement:

$$\theta=\frac{kp}{1+kp} \tag{3.27}$$

or

$$\alpha^{s}=\frac{\alpha_m^{s}kp}{1+kp}. \tag{3.28}$$

Taking account of equation (3.20) we can also write:

$$a=\frac{a_m kp}{1+kp}. \tag{3.29}$$

Equations (3.27), (3.28) and (3.29) represent different forms of the Langmuir adsorption isotherm, describing localized adsorption on a homogeneous surface when no interactions take place between the adsorbate molecules. Langmuir derived his adsorption isotherm in kinetic terms, but it can also be derived thermodynamically [15] or statistically [16].

For low pressures in the gas phase, $kp \ll 1$, and

$$\theta \approx kp, \tag{3.30}$$

$$a = a_{\mathrm{m}} kp, \tag{3.31}$$

$$\alpha^{\mathrm{s}} = \alpha^{\mathrm{s}}_{\mathrm{m}} kp \tag{3.32}$$

which shows that adsorption is proportional to p, and hence, in this pressure range, the Langmuir equation reduces to Henry's equation.

If the gas phase adsorbate pressure is sufficiently high, then $kp \gg 1$, and we can neglect unity in the denominator of equations (3.27), (3.28) and (3.29), yielding:

$$\theta \to 1, \quad a \to a_{\mathrm{m}}, \quad \alpha^{\mathrm{s}} \to \alpha^{\mathrm{s}}_{\mathrm{m}}.$$

Hence adsorption initially increases linearly with pressure, after which this growth gradually decreases, and at sufficiently high gas pressures adsorption assumes a constant value: the adsorbent surface becomes saturated with a monomolecular adsorbate layer (the Langmuir adsorption isotherm is shown in Fig. 3.3).

The Langmuir adsorption isotherm, e.g. equation (3.29), can be written in the form of linear equations:

$$\frac{p}{a} = \frac{1}{a_{\mathrm{m}} k} + \frac{1}{a_{\mathrm{m}}} p \tag{3.33}$$

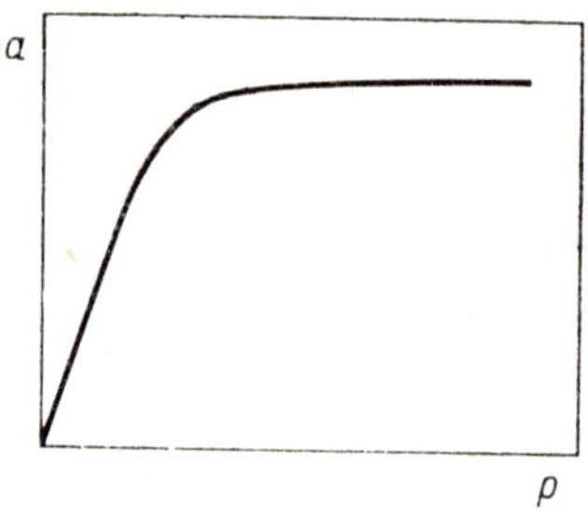

FIG. 3.3 Langmuir's adsorption isotherm.

or

$$\frac{a}{p}=ka_{\mathrm{m}}-ka \tag{3.34}$$

or

$$\frac{1}{a}=\frac{1}{a_{\mathrm{m}}}+\frac{1}{a_{\mathrm{m}}k}\frac{1}{p}\,. \tag{3.35}$$

The plots of p/a versus p, a/p versus a or $1/a$ versus $1/p$ are, when Langmuir's equation is applicable, straight lines, permitting the calculation of the constants a_{m} and k.

The quantitity a_{m}, i.e. the amount of adsorbate (mole/g) covering the surface area of the adsorbent in a monomolecular film, is known as the monolayer capacity. This quantity allows us to calculate the specific surface area of the adsorbent, if we know the surface ω_{m} occupied by a molecule in the monolayer (the sitting surface),

$$s=a_{\mathrm{m}}N\omega_{\mathrm{m}} \tag{3.36}$$

where N is the Avogadro number.

The Langmuir adsorption isotherm has a simple form, an equally simple physical picture, and yields reasonable agreement with a large number of experimental isotherms. Thus the Langmuir adsorption isotherm is probably the best known of all isotherms describing adsorption, and often serves as a basis for more detailed developments [17].

Although Langmuir's equation often allows a satisfactory interpretation of adsorption data, more critical studies reveal various discrepancies due to the occurrence of multilayer adsorption with lowering of temperature and heterogeneity of adsorbent surface. In general, the Langmuir isotherm holds best in the cases of chemisorption and adsorption from solution of comparatively large molecules such as dyes.

In adsorption of gases from a mixture, separate adsorption equilibria exist between the components of the mixture and the free surface of the adsorbent. The adsorption isotherm for component i (fraction of surface coverage by molecules of component i) then has the form:

$$\theta_i=\frac{k_i p_i}{1+\sum\limits_i k_i p_i} \tag{3.37}$$

where p_i are the partial pressures of the mixture components, and k_i are the corresponding adsorption constants. Thus adsorption of a given component of a multicomponent gas mixture increases with increase of its partial pressure and with decrease of partial pressures of the remaining components.

3.4.4 Potential Theory of Adsorption

Langmuir's theory is based on the fundamental assumption that the adsorption layer is monomolecular, when, even at complete covering of the adsorbent surface, the number of molecules adsorbed cannot exceed the number of active sites. The adsorption monolayer completely shields the action of adsorption forces and thus inhibits the formation of the next layer.

Although Langmuir's theory has found support in many instances, especially in measurements of gas adsorption at low pressures, nevertheless it does not always adequately explain observed phenomena. Thus, alongside Langmuir's theory, a theory of a multilayer adsorption (potential theory of adsorption) was advanced by Eucken [18] and Polanyi [19]. The latter assumed that adsorption forces act at distances greatly exceeding the dimensions of a single molecule and that they are not totally shielded by the first layer of adsorbate. Under these conditions the adsorption layer has a diffusional character and its density varies with distance from the adsorbent surface (cf. the variation of atmospheric density with distance from the earth's surface). The potential theory depends upon the absorption potential ε and the volume of the adsorbed layer V^s.

The adsorption potential corresponds to the change of molar free energy connected with the change of vapour pressure from that over the pure liquid phase p_0 to equilibrium pressure p at a given coverage of the adsorbent surface,

$$\varepsilon = RT \ln \frac{p_0}{p}. \qquad (3.38)$$

At the adsorbent surface there is a force field known as the adsorption potential field. In the surface space we can draw the anticipated equipotential surface (of equal adsorption potential). The cross-section of the surface layer is shown in Fig. 3.4. Equipotential surfaces are represented in the form of dashed lines. The spaces between every two equipotential surfaces have definite volumes. There is therefore a relationship between the adsorption potential and the surface layer volume V^s:

$$\varepsilon = f(V^s), \tag{3.39}$$

$$V^s = aV_m \tag{3.40}$$

where a is the adsorption (in moles per gram of adsorbent), and V_m is the molar volume of liquid whose vapour is adsorbed (at the temperature of the experiment).

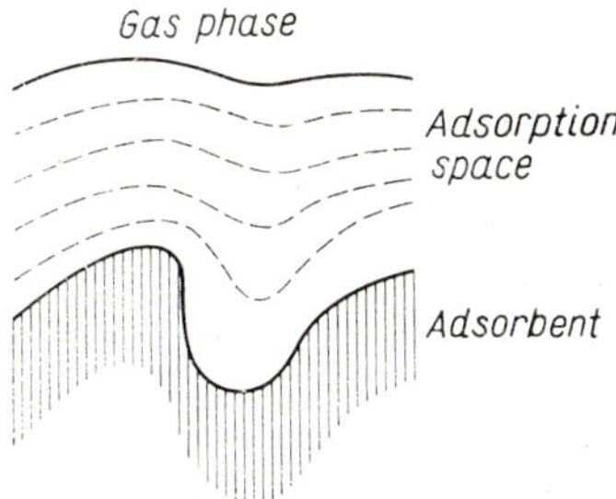

FIG. 3.4 Cross-section of the surface layer in terms of the potential theory (after Brunauer [1]).

With increasing distance from the surface we observe an increase of the volume of the surface layer and a decrease of the adsorption potential.

Figure 3.5 represents the adsorption potential as a function of surface layer volume V^s. Polanyi named that curve the characteristic adsorption curve, and the function $\varepsilon = f(V^s)$ the characteristic adsorption function.

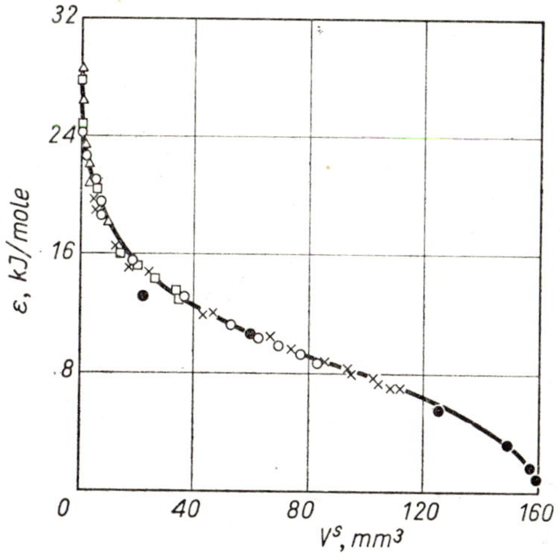

FIG. 3.5 Characteristic curve of CO_2 adsorption on carbon (after Brunauer [1]).

Polanyi assumed that the adsorption potential is temperature independent over a wide range of temperature:

$$\left(\frac{\partial \varepsilon}{\partial T}\right)_{V^s}=0 \tag{3.41}$$

or

$$\varepsilon=RT_1 \ln \frac{p_{0,1}}{p_1}=RT_2 \ln \frac{p_{0,2}}{p_2}. \tag{3.42}$$

Thus the characteristic adsorption curve is temperature-invariant (Fig. 3.5).

The potential theory does not yield a definite adsorption isotherm. This isotherm is replaced to some extent by the characteristic adsorption curve. Experimental testing of the theory consists in calculating the characteristic curve on the basis of an experimental isotherm and hence determining the isotherms at different temperatures.

The volume of the adsorbed layer V^s, its density ρ_s and amount of adsorbed substance a are related by:

$$V^s=a/\rho_s.$$

If we determine experimentally the values a and p, and know the values p_s and ρ_s for the gas under investigation at the given temperature, we can express ε as a function of V^s and thus determine the adsorption isotherms at other temperatures. The investigations of Titoff [20] and the calculations of Berenyi [21] have shown that the isotherms for CO_2 adsorption on activated carbon obtained in this way are in excellent agreement with experiment (Fig. 3.6).

For equal distances from the adsorbent surface, and hence equal values of V^s, the ratio of adsorption potentials for two different adsorbed vapours and one given adsorbent is constant:

$$\left(\frac{\varepsilon}{\varepsilon_0}\right)_{V^s}=\beta_a \tag{3.43}$$

where ε_0 is the adsorption potential for vapour of a standard adsorbate, and the coefficient β_a is the convergence (affinity) coefficient of the characteristic curves.

One further feature of Polanyi's theory deserves attention. If we assume that the state of the adsorption layer can be described by the van der Waals equation, then at sufficiently low temperature it will be possible to observe not only an increase of the gas concentration but also its condensation on the adsorbent surface proper. We should distinguish such a condensation, which can proceed only on the flat surface of the adsorbent, from capillary condensation, which takes place only if the adsorbent has a porous structure.

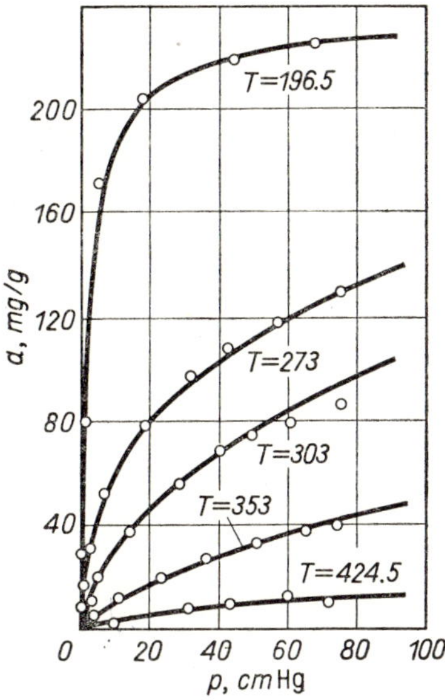

FIG. 3.6 Isotherms of CO_2 adsorption on carbon (after Brunauer [1]): empty circles denote Titoff's experimental points, solid lines represent the curves calculated in terms of the potential by Berenyi.

Dubinin *et al.* [22–24] as well as Radushkevich [25] have found that the characteristic adsorption curve is related to the porous structure of the adsorbent. For activated carbons with narrow pores (of 1st structural type) the applicability of relation (3.43) has been repeatedly confirmed.

An attempt to provide theoretical justification for the characteristic curves for microporous adsorbents produced:

$$V^s = V_0^s e^{-k\varepsilon^2/\beta_a^2} \tag{3.44}$$

where V_0^s is the so-called limiting volume at potential $\varepsilon = 0$, approximately equal to the volume of the micropores; and k is the parameter expressing the pores' volume distribution function by their dimensions. Taking account of equation (3.38) and putting $b = k/\beta_a^2$, we can write the Dubinin and Radushkevich adsorption isotherm (D–R equation):

$$V^s = V_0^s \, e^{-b\left(RT \ln \frac{p_0}{p}\right)^2}. \tag{3.45}$$

Dividing both sides of this equation by the molar volume of the liquid adsorbate,

$$a = a_0 \, e^{-b\left(RT \ln \frac{p_0}{p}\right)^2} \tag{3.46}$$

where a_0 is the number of moles of liquid adsorbate required to fill the micropores of 1 g of the adsorbent. The latter equation can easily be rewritten by the substitution:

$$D = \frac{0.4343BT^2}{\beta_a^2} \tag{3.47}$$

where

$$B = 2.303^2 \kappa R^2. \tag{3.48}$$

Then

$$a = a_0 \times 10^{-D\left(\log \frac{p_0}{p}\right)^2} \tag{3.49}$$

and taking logarithms we obtain

$$\log a = \log a_0 - D\left(\log \frac{p_0}{p}\right)^2 \tag{3.50}$$

and for V^s

$$\log V^s = \log V_0^s - D\left(\log \frac{p_0}{p}\right)^2. \tag{3.51}$$

If the experimental data are represented in the form of $\log a$ or $\log V^s$ versus $(\log(p_0/p))^2$ plots, we obtain a straight line from which we can find $\log a_0$ or $\log V_0^s$ and D. It has been shown that the values of the constant B in the D–R adsorption isotherm are directly associated with the porous structure of the adsorbent.

In Fig. 3.7 we illustrate nitrogen vapour adsorption isotherms for several different adsorbents at 67 K in the D–R equation co-ordinate system [26]. As can be seen, the isotherms are characterized by long rectilinear segments encompassing only relative low equilibrium pressure of the adsorbate (for p/p_0 values ranging from 0 to about 0.01). At higher adsorbate pressures, the adsorption isotherms strongly deviate from linearity. In order to obtain the value of $\log a_0$ (or $\log V_0^s$) the rectilinear

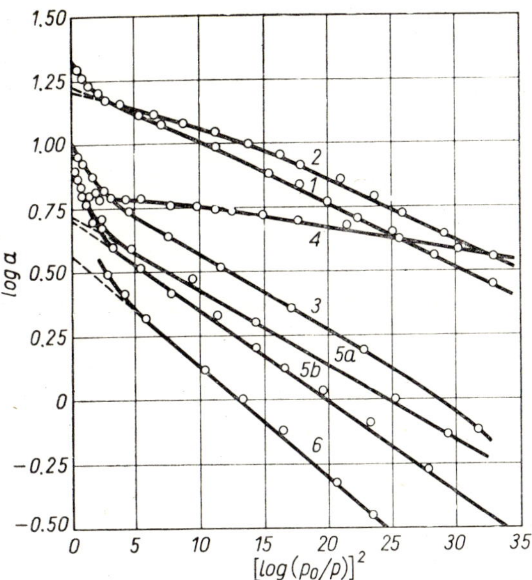

Fig. 3.7 Nitrogen vapour adsorption isotherms on several adsorbents at 67 K as plotted in the D–R co-ordinate system (after Dubinin and Zaverina [26]): *1*—activated carbon from coconut shells, *2*—SKT activated carbon, *3*—silica gel S-200, *4*—chabazite, *5a*—silica gel S-204 (90 K), *5b*—silica aerogel. Adsorption a is given in mmole/g.

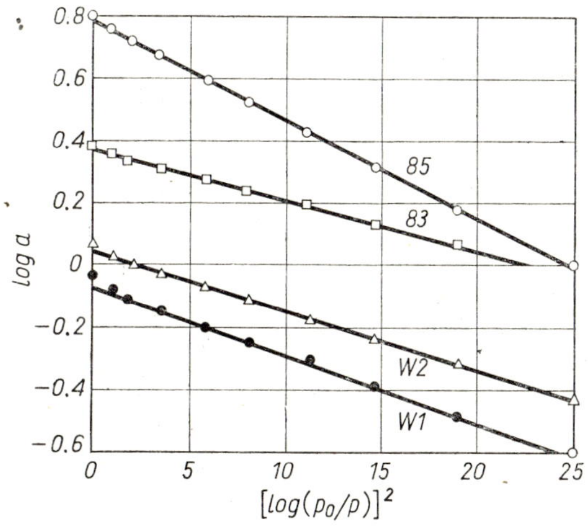

Fig. 3.8 Benzene vapour adsorption isotherms on several kinds of activated carbon at 293 K. Adsorption a is given in mmole/g (after Dubinin and Zaverina [26]).

segment of the isotherm should extrapolated to relative pressure $p/p_0=1$ (when $\log(p_0/p)=0$). In Fig. 3.8 we observe adsorption isotherms at 293 K for benzene vapour on several kinds of activated carbon obtained from sugar. They are also plotted in the D–R co-ordinate system. Agreement between experiment and theory is very good since the adsorption isotherms are linear in this co-ordinate system over the whole range of adsorbate relative pressures.

3.4.5 The Brunauer, Emmett and Teller Theory of Multilayer Vapour Adsorption

Vapour adsorption is characterized by condensation at saturated vapour pressure, when the adsorption of vapour of a liquid wetting the adsorbent surface becomes infinitely high. Therefore if, in the zone where the monolayer is being filled up, extent of adsorption decreases with in-

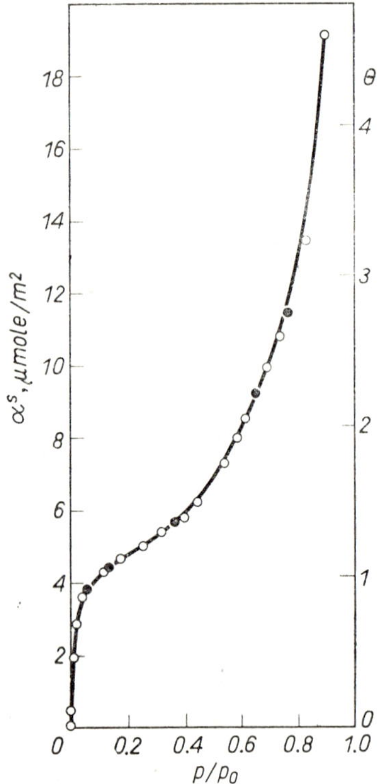

Fig. 3.9 Adsorption–desorption isotherm for benzene vapour on graphitized carbon black (after [15]).

creasing pressure (convex sector of the isotherm), further increase of pressure, when p tends to p_0, should increase the adsorption. In this case the adsorption layer becomes thicker, and, at $p=p_0$, volume condensation takes place. Adsorption then becomes multilayer, and the isotherm passes through a point of inflexion. In Fig. 3.9 the isotherm is shown for benzene adsorption at 273 K on a homogeneous graphitized carbon black surface. The isotherm is S-shaped and fully reversible, i.e. desorption proceeds along the same curve. Such an adsorption isotherm shape is neither predicted by the Langmuir theory nor accurately described in terms of the Polanyi theory.

In 1938 Brunauer, Emmett and Teller [27] developed an interesting multilayer adsorption theory (BET theory), based on Langmuir's theory. The principal assumption of the BET theory is that the Langmuir equation applies to every adsorption layer: a molecule encountering an occupied site on the adsorbent surface does not leave that site immediately, but forms a short-lived adsorption complex. With increasing vapour pressure, when p approximates the saturated vapour pressure p_0, the number of free sites on the adsorbent surface decreases, and therefore the number of active sites occupied by single adsorbate molecules decreases as well since double, triple, etc. adsorption complexes are formed (see Fig. 3.10)

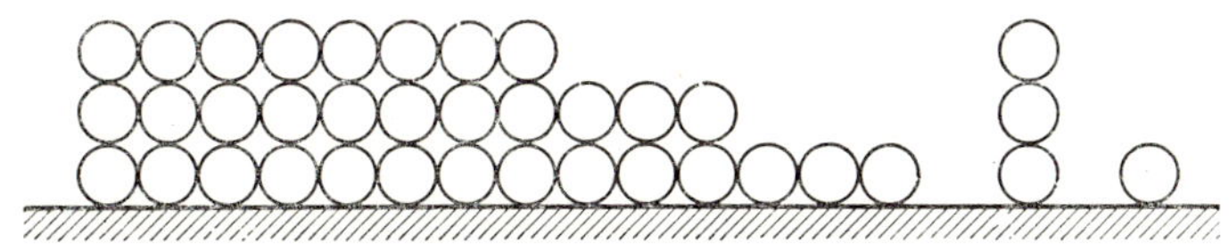

FIG. 3.10 Multilayer model of BET adsorption.

Brunauer, Emmett and Teller have considered (like Langmuir) the adsorption process in terms of kinetic theory and derived a multilayer adsorption isotherm, which can also be derived statistically [28].

In the following, this equation is derived by an analysis of adsorption equilibrium. If we disregard interactions between adsorbate molecules along the adsorbent surface, we can consider multilayer adsorption of vapour on a homogeneous surface as a series of equilibrium steps:

vapour + free surface area ⇄ single complex,
vapour + single complex ⇄ double complex,
vapour + double complex ⇄ triple complex, etc.

Let $\theta', \theta'', \theta''', \ldots$ be the fractions of adsorbent surface area covered with single, double, triple, etc. adsorption complexes. The total amount of adsorbed vapour is then given by:

$$a = a_m(\theta' + 2\theta'' + 3\theta''' + \ldots) \tag{3.52}$$

where a_m is the monolayer capacity.

The various equilibrium constants are:

$$k' = \frac{\theta'}{p\theta_0}, \quad k'' = \frac{\theta''}{p\theta'}, \quad k''' = \frac{\theta'''}{p\theta''}, \ldots \tag{3.53}$$

where θ_0 is the fraction of adsorbent area not occupied by adsorbate molecules. The value of k' is usually much higher than that of k'', since the adsorbent–adsorbate interaction rapidly decreases with increasing distance from the surface. Constants $k'', k''', \ldots$ are not equal, but their differences are usually much smaller than that between k' and k''. For this reason, the BET theory assumes that:

$$k'' \approx k''' \approx \ldots \approx k_L \tag{3.54}$$

where k_L is the equilibrium constant for the saturated vapour $\rightleftarrows$ liquid system (the condensation constant) and equals $1/p_0$ (if for the vaporization equilibrium: liquid $\rightleftarrows$ saturated vapour, $k_p = p_0$, then for the condensation equilibrium $k_L = 1/k_p$). Hence from equation (3.53) we get:

$$\theta' = k' p\theta_0, \tag{3.55}$$

$$\theta'' = k'' p\theta' = k_L\, p\theta' = \frac{p}{p_0}\,\theta', \tag{3.56}$$

$$\theta''' = k''' p\theta'' = (k_L\, p)^2\theta' = \left(\frac{p}{p_0}\right)^2 \theta', \quad \text{etc.} \tag{3.57}$$

Substituting expressions (3.55), (3.56) and (3.57) into equation (3.52) we get:

$$a = a_m k' p\theta_0 \left[1 + 2\frac{p}{p_0} + 3\left(\frac{p}{p_0}\right)^2 + \ldots\right]. \tag{3.58}$$

But:

$$\theta_0 + \theta' + \theta'' + \theta''' + \ldots = \theta_0 \left\{1 + k'p\left[1 + \frac{p}{p_0} + \left(\frac{p}{p_0}\right)^2 + \ldots\right]\right\} = 1. \tag{3.59}$$

Since $p/p_0 \leqslant 1$, the sum of the geometrical series in square brackets in equation (3.58) is equal to $(1 - p/p_0)^{-2}$ and that in equation (3.59) is equal to $(1-p/p_0)^{-1}$. Hence we can write equation (3.58) in the form:

$$a=\frac{a_m k' p}{(1-p/p_0)(1+k'p-p/p_0)} . \tag{3.60}$$

Taking into account that

$$p=p_0\frac{p}{p_0}=\frac{1}{k_L}\frac{p}{p_0} ,$$

and putting

$$k'/k_L=C ,$$

we finally get:

$$a=\frac{a_m C\, p/p_0}{(1-p/p_0)[1+(C-1)\,p/p_0]} \tag{3.61}$$

or, analogously:

$$\theta=\frac{C\, p/p_0}{(1-p/p_0)[1+(C-1)\,p/p_0]} . \tag{3.62}$$

Equation (3.61) or (3.62) is the Brunauer, Emmett and Teller (BET) isotherm equation of multilayer vapour adsorption. The constant $C=k'/k_L$ is associated with the difference between the enthalpy of adsorption of the first layer and the enthalpy of condensation. We can write (see Section 3.5) that:

$$C=g_0 \exp\left(-\frac{\Delta_a H_m^s-\Delta H_m^L}{RT}\right) \tag{3.63}$$

where $\Delta_a H_m^s$ is the molar enthalpy of first layer adsorption, ΔH_m^L is the molar enthalpy of condensation, and $g_0=\exp(\Delta_a S_m^s-\Delta S_m^L)/R$ is the so-called entropic factor. The difference $\Delta_a H_m^s-\Delta H_m^L$ is referred to as the pure enthalpy of adsorption.

The BET equation has filled a gap in adsorption theory and has been accepted as the general method for determining the adsorbent surface area from adsorption data, since equation (3.61) can be written in the linear form:

$$\frac{p/p_0}{a(1-p/p_0)}=\frac{1}{a_m C}+\frac{C-1}{a_m C}\frac{p}{p_0} . \tag{3.64}$$

If we represent the adsorption isotherm in the $(p/p_0)/a(1-p/p_0)$ and p/p_0 co-ordinate system, constants a_m and C can be determined from the slope of the straight line and its point of intersection with the ordinate (Fig.

3.11), namely, tan $\alpha=(C-1)/a_m C$, and section $b=1/a_m C$. The specific surface area of the adsorbent can be calculated from equation (3.36) for known a_m.

Conventional measurements of adsorbent specific surface area by the BET method are performed with the use of the low-temperature (78 K) nitrogen adsorption isotherm, assuming that $\omega_m=0.162\ nm^2$.

According to Brunauer [1] five principal forms of adsorption isotherm for gases and vapours can be distinguished (Fig. 3.12).

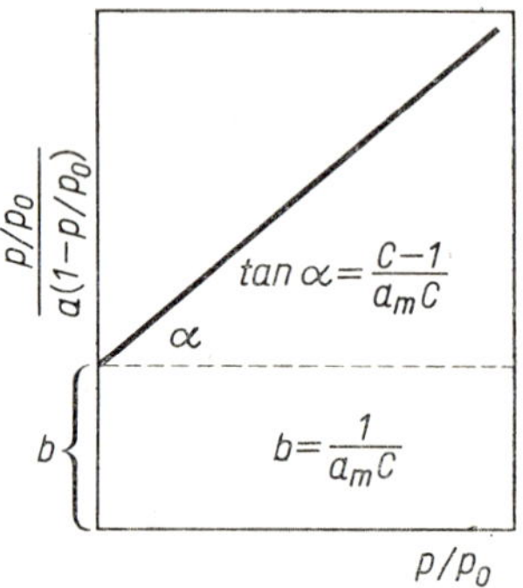

FIG. 3.11 Determination of the constants in the BET adsorption isotherm.

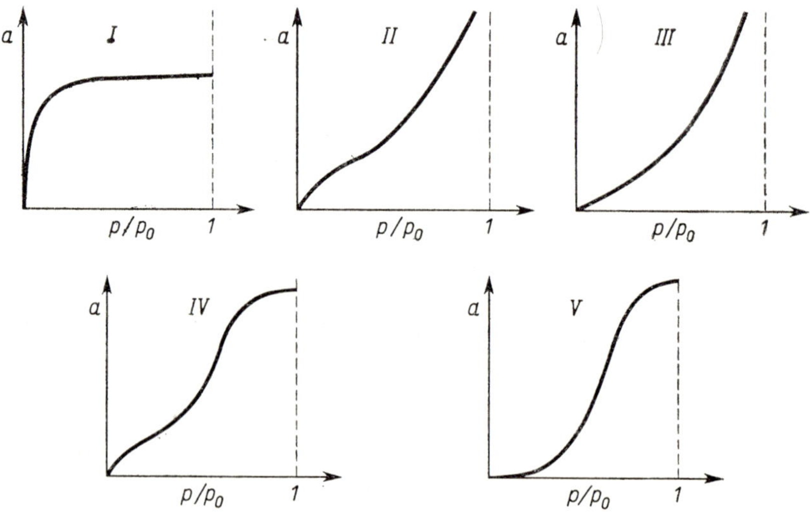

FIG. 3.12 Types of gas and vapour adsorption isotherms according to Brunauer [1].

Type I corresponds to the Langmuir isotherm and is characterized by the fact that it tends monotonically to the limiting adsorption, probably attained on completion of the monomolecular layer.

Type II is very common in cases of physical adsorption, and is probably associated with the formation of a multimolecular adsorption layer; see Fig. 3.9 for adsorption of benzene vapour on graphitized carbon black.

Type III is rather uncommon, but is exemplified by adsorption of bromine on silica gel. It seems that in this case the absolute value of the heat of adsorption is equal to or smaller than the heat of condensation of the pure adsorbate.

Isotherms of Types IV and V correspond to those of Types II and III, with the difference that the adsorption maximum is reached at a pressure lower than that of the saturated vapour p_0. It is believed that they both reflect the phenomenon of capillary condensation.

The BET equation encompasses the first three such adsorption isotherm types. When $|\Delta_a H_m^s| \gg |\Delta H_m^L|$, C is very large, so equation (3.61) reduces to:

$$a = \frac{a_m C p/p_0}{1 + C p/p_0} \tag{3.65}$$

This is the Langmuir isotherm, i.e. Type I isotherm. For values of C in the range 3 or 4 to several hundred, the BET equation yields isotherms corresponding to Type II. If C is equal to or smaller than unity, i.e. if $|\Delta_a H_m^s| \ll |\Delta H_m^L|$, the BET equation yields, according to Brunauer, isotherms

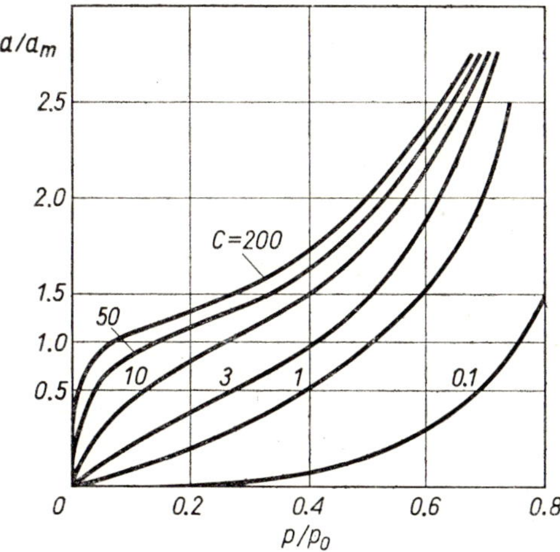

FIG. 3.13 The BET adsorption isotherms for various values of C (after Brunauer [1]).

of Type III. Such isotherms have been observed for halogen adsorption on carbon and silica gel as well as for adsorption of water and organic vapours on glass. In Fig. 3.13, BET isotherms are plotted for various values of C.

The BET adsorption isotherm is convenient for experimental use because it can be represented in a linear form and requires the selection of only two parameters. Agreement of this equation with experimental data lies, however, in a rather narrow range. For instance in the case of adsorption isotherms of Type II, plotted according to the BET equation in linear form, the straight-line segment is obtained for the range of p/p_0 values of 0.05–0.3. In the range 0.3 to 0.5 the BET equation usually does not hold, because the difference between the constants k'', k''', etc. plays a significant role which cannot be neglected. The adsorption predicted by the BET equation is usually too small at low pressures and too large at high pressures.

If, due to limiting of the surface layer (e.g in a capillary), only n layers (instead of the previous infinite number) are formed in the adsorption process, then the BET isotherm has the form:

$$a=\frac{a_m Ch[1-(n+1)h^n+nh^{n+1}]}{(1-h)[1+(C-1)h-Ch^{n+1}]} \tag{3.66}$$

where $h=p/p_0$. When $n=1$, equation (3.66) becomes, for all values of C, the Langmuir equation with $C/p_0=k$. If adsorption takes place on the free surface, then $n\to\infty$ and equation (3.66) reduces to equation (3.61).

There exist several further modifications of the BET equation [29, 30], of which the Hüttig equation deserves particular mention [31, 32]. Hüttig assumes that adsorbate molecules can undergo desorption even if covered by a second layer, and the Hüttig adsorption isotherm can be written in the form:

$$a=\frac{a_m Ch(1+h)}{1+Ch} \tag{3.67}$$

or, in the linear form:

$$\frac{h(1+h)}{a}=\frac{1}{Ca_m}+\frac{1}{a_m}h. \tag{3.68}$$

The BET and Hüttig equations become equivalent as h approaches zero. At higher equilibrium pressures, the BET equation predicts excessive gas adsorption, while the values found from the Hüttig equation are too low. Equation (3.67) may be considered as the Langmuir adsorption isotherm multiplied by the factor $(1+h)$ to allow for adsorption in the second and further layers. If we assume complete filling of the monolayers, the coefficient $(1+h)$ acts in such a way that the adsorption isotherm is a straight line above the inflexion point (for $a \approx a_m$), in agreement with experiment.

In general, however, modifications of the BET equation, both those described here and others, have not found wide acceptance by those studying gas or vapour adsorption processes.

3.4.6 Interaction of Molecules in the Adsorption Layer

The Henry, Langmuir and BET adsorption isotherms neglect "horizontal" interactions between adsorbed molecules in the surface layer. The BET theory allows for interactions between adsorbate molecules solely in the direction vertical to the adsorbent surface, and the potential theory of adsorption does not in fact distinguish between interactions of gas or vapour molecules in the bulk phase and in the adsorption layer.

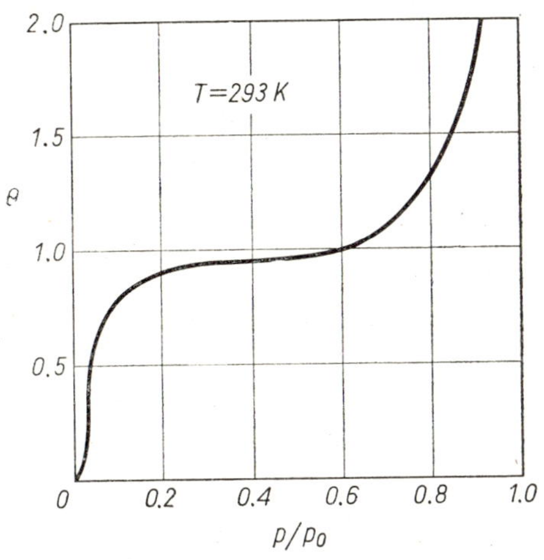

Fig. 3.14 Carbon tetrachloride adsorption isotherm on graphitized carbon black at 293 K.

The effect of such horizontal interactions on the adsorption process may often be quite significant, for instance, in the course of adsorption, on a non-polar surface, of large molecules (e.g. CCl_4, $C(CH_3)_4$) or of molecules forming associated compounds via hydrogen bonds on the adsorbent surface. Adsorbate–adsorbate interactions (horizontal interactions) result in bending of the adsorption isotherm towards the pressure axis (Fig. 3.14) and an increase in the heat of adsorption on homogeneous adsorbent surface with coverage (see Section 3.5).

The derivation of an adsorption isotherm which provides for all possible intermolecular interactions in the adsorption layer must encounter great difficulties. Therefore the simple solution advanced by Kiselev [33] for the case of vapour adsorption on homogeneous surface, e.g. graphitized carbon black, deserves special attention. He assumed two quasichemical equilibria for monomolecular adsorption layers:

(i) gas molecule + free surface ⇄ single adsorption complex,

(ii) { single complex + single complex ⇄ double horizontal adsorption complex,
single complex + double complex ⇄ triple horizontal adsorption complex.

Kiselev then derived the approximate adsorption equilibrium equation:

$$h = \frac{\theta'}{K_1'(1-\theta')(1+K_n\theta')}. \qquad (3.69)$$

If we put $K_1 = K_1'/p_0$, then

$$p = \frac{\theta'}{K_1(1-\theta')(1+K_n\theta')} \qquad (3.70)$$

where θ' is the monomolecular coverage of the adsorbent surface, K_1 is the equilibrium constant for reaction (i) (vertical adsorbate–adsorbent interaction), and K_n is the equilibrium constant for reaction (ii) (horizontal adsorbate–adsorbate interaction). Equation (3.69) or (3.70) reduces to the Langmuir adsorption isotherm when $K_n = 0$.

For poorly adsorbing substances, when multilayer gas adsorption is taken into account, we should consider the additional equilibrium:

(iii) gas molecule + horizontal complex ⇄ double vertical adsorption complex.

This yields an approximate multilayer adsorption equation which allows for horizontal interactions in the first adsorbate layer:

$$h = \frac{\theta'(1-h)^2}{K_1'[1-\theta(1-h)][1+K_n\theta(1-h)]}. \qquad (3.71)$$

When $K_n = 0$ equation (3.71) has the form of the BET equation.

Experimental results for vapour adsorption on graphitized carbon black correspond to Kiselev equations, since adsorption takes place on practically homogeneous surfaces, when other effects do not interfere with the horizontal adsorbate–adsorbate interaction (Fig. 3.15 and Table 3.1).

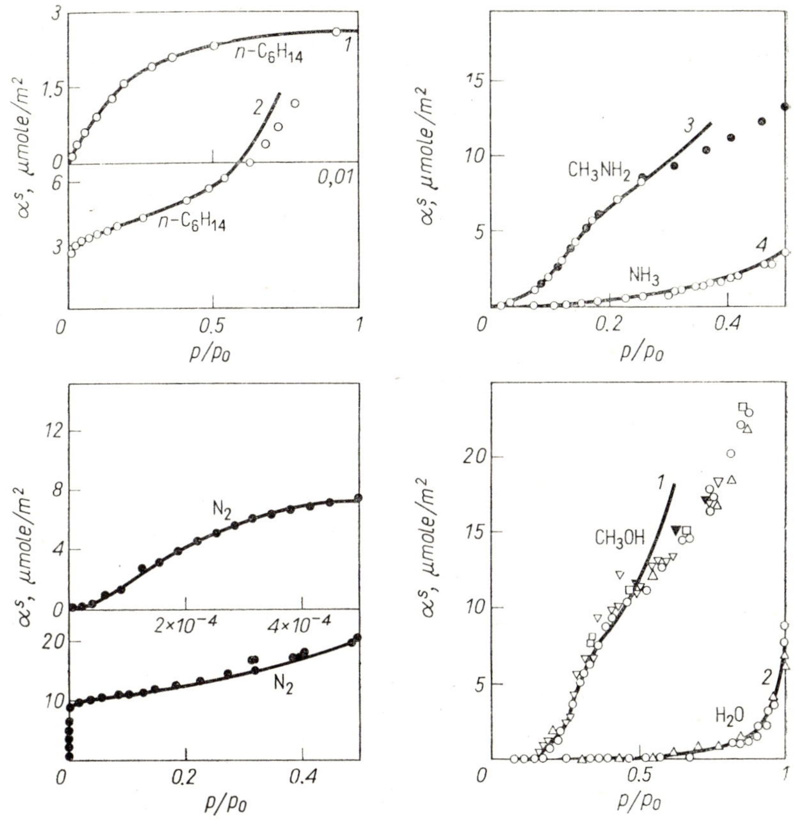

FIG 3.15 Adsorption isotherms for vapours of several substances on graphitized carbon black (after Kiselev [33]).

The Kiselev equations only describe localized adsorption, and so the attempt to include horizontal interactions in the non-localized monolayer described by Hill [34] and De Boer [35] deserves mention. They derived an adsorption isotherm for the homogeneous surface in the form:

$$h=\frac{\theta}{K_1(1-\theta)}\exp\left(\frac{\theta}{1-\theta}-K_2\theta\right) \tag{3.72}$$

Table 3.1 EQUATION (3.71) CONSTANTS FOR ADSORPTION ON GRAPHITIZED CARBON BLACK (AFTER KISELEV [33])

Adsorbate	Temperature K	Monolayer capacity	K_1'	K_n	$\frac{K_n}{K_1'}$
n-Hexane	293	3.08	500	0.5	0.001
Nitrogen	78	10.3	1000	6	0.006
Methylamine	273	10.4	0.75	9.5	13
Ammonia	194.2	14.6	0.094	4.6	50
Methanol	293	10.4	0.10	35	350
Water	303	15.6	(0.005)	(250)	(50 000)

where $K_2 = a_2/kTb_2$ (a_2 and b_2 being constants of the two-dimensional van der Waals equation). Hill [36] has also proposed a multilayer adsorption isotherm which accounts for horizontal interactions in the surface layer.

3.4.7 THE HARKINS–JURA EQUATION

Harkins and Jura [37, 38] proposed an adsorption isotherm based on the formation of an adsorbate film on the adsorbent surface, and obtained the expression:

$$\log \frac{p}{p_0} = B - \frac{A}{v^2} \tag{3.73}$$

where A and B are constants, and v is the volume of adsorbed vapour. Equation (3.73), the Harkins–Jura adsorption equation (H–J equation), shows that the $\log p/p_0$ versus $1/v^2$ plot should be a straight line. Adsorption isotherms of Type II are often in agreement with that equation (Fig. 3.16b). The slope of the straight line gives the value of the constant A, which is related to the specific surface area of the adsorbent by:

$$s = kA^{1/2}. \tag{3.74}$$

For nitrogen adsorption at 77.2 K, when s is expressed in m^2/g, we have $k = 4.06$.

The H–J equation yields adsorption isotherms of Type II which, in general, correspond to BET isotherms. The authors have left open the problem of the significance of the variable value of constant B. The H–J isotherms with various values of constant B are shown in Fig. 3.16a. The

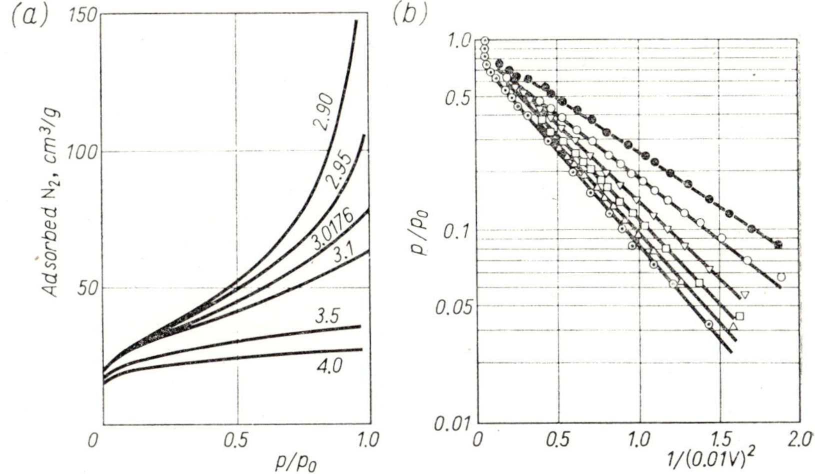

Fig. 3.16 (a) Harkins–Jura adsorption isotherms for various values of constant B, (b) H–J plots of N_2 adsorption on various porous adsorbents. Specific surface areas of the solid adsorbent in m^2/g: ● 321, ○ 365, ▽ 395, □ 401, △ 438, ⊙ 455 (after Harkins and Jura [37]).

range of agreement of the H–J equation with experiment is usually greater than that for BET equations. Sometimes, the H–J plot consists of two straight lines, in which case the authors recommend making use of the slope of the low-pressure part.

3.4.8 Comparison of Gas and Vapour Adsorption Theories

Of the various gas and vapour adsorption theories the Polanyi potential theory is still considered to be valid, although sixty years have elapsed since its publication, on account of its thermodynamic character. This theory does not define a specific adsorption isotherm nor does it give a detailed mechanism. Extention of the potential theory of adsorption by Dubinin and Radushkevich for microporous adsorbents has led to a further adsorption isotherm (the D–R equation). The introduction of additional assumptions into the Polanyi theory, however, affects its purely thermodynamic character.

The Langmuir and BET theories introduce the concept of localized mono- and multilayer adsorption of gases (vapours). This concept has been subjected to thermodynamic verification [39]. The least we demand from every theory is that it accurately describes the free energy change. This is related to differing enthalpy and entropy effects at varying degrees of coverage of the adsorbent surface. The enthalpy and entropy contributions arising from the BET theory agree at best semiquantitatively with experiment. The calculated enthalpy values are too small and the entropy values too high. There are therefore serious doubts as to the applicability of the physical picture given by the BET theory (and also that of Langmuir).

Despite the above theoretical deficiencies, the BET adsorption isotherm has been generally accepted as a basis for the standard method of determining the specific surface area of adsorbents (low-temperature adsorption of nitrogen vapour); this is due to its convenience as well as to its special position in the family of adsorption isotherms. For solids the specific surface area as determined by the BET equation is the surface area of the adsorbent together with irregularities of dimensions greater than or comparable to the dimensions of the adsorbate molecules. Consistent relative values of specific surface area have been obtained.

Some authors [40–42] draw attention to the fact that the BET and H–J equations show convergence in the sense that if a set of data satisfy one of the equations it will also satisfy the other, especially for adsorption isotherms in which the value of the BET constant C is 50 or more. There are many points, however, at which the two theories are incompatible. It is believed that the two-dimensional adsorbate film on the adsorbent surface as assumed by the H–J theory is probably one of the better models in the monolayer formation region; this model seems indispensable for explaining two-dimensional condensation phenomena. For p/p_0 values greater than 0.1 the properties of the adsorbent film tend to those of a normal liquid.

The above comparison demonstrates that the extension of the theory of gas and vapour adsorption to cover the whole range of pressures from $p/p_0=0$ to $p/p_0=1$, has not as yet been successful. Such studies are still continuing.

Ten years ago Jovanovič [43, 44] derived an isotherm for both mono- and multilayer gas adsorption. For monolayer physical adsorption, the Jovanovič equation has the form:

$$a=a_m[1-\exp(-a''h)]. \tag{3.75}$$

For multilayer adsorption this equation becomes:

$$a=a_m[1-\exp(-a''h)]\exp(b''h) \tag{3.76}$$

where h is the relative equilibrium pressure of the gas, a'' and b'' are constants which describe respectively adsorption in the first and successive layers of the adsorbed substance:

$$a''=\tau_1 B \quad \text{and} \quad b''=\tau_L B$$

where again $B=\omega_m p_0(2\pi\, mkT)^{1/2}$, τ_1 and τ_2 are the residence times of gas molecules in the first and further layers, ω_m is the sitting surface of deposition of gas molecules, m is the mass of the gas molecule, and k is the Boltzmann constant.

Based on this equation, Jovanovič distinguished as many as eight types of gas and vapour adsorption isotherms.

Further studies are expected to lead to even better results, providing a more accurate description of the true mechanism of adsorption from the gaseous phase.

3.4.9 Effect of Capillary Dimensions on the Adsorption of Vapours at Low Relative Pressures

When the capillaries in the adsorbent become narrow, the adsorption forces exerted by their walls may reinforce and the adsorption energy may increase. This increase will be particularly significant for highly polarizable adsorbate molecules, e.g. large hydrocarbon molecules and their derivatives. For instance, the adsorption energy of hexane or benzene vapour on silica gel increases significantly in adsorbent capillaries of diameters 4–5 nm.

When small molecules, e.g. nitrogen or methanol, are involved the adsorption energy already changes significantly at pore diameters smaller than 3 nm. For still smaller molecules, e.g. water, narrowing of the capillaries to 2.5 nm does not affect the adsorption energy.

In the case of adsorbents with narrow pores, we cannot speak of a uniform filling of the adsorption layer, as distinguished from adsorption on non-porous surfaces. The adsorption potential is higher in narrow as distinct from wide capillaries. In narrow capillaries, strong adsorption occurs and the adsorbate concentration is higher than on the surface of wide capillaries. The application of both the Langmuir and BET equations to adsorbents with a partly heterogeneous surface becomes difficult. Even if it were possible to describe adsorption on adsorbents with narrow pores using these equations, the constant a_m loses its physical significance, and the use of that constant to determine the specific surface area of such adsorbents does not yield correct results.

3.4.10 Effect of Capillary Dimensions on the Adsorption of Vapours at High Relative Pressures. Capillary Condensation

In the case of porous adsorbents, adsorption proceeds at low p/p_0 values on the surface of capillaries exactly as with non-porous materials, except for their zones of strong narrowing. At elevated vapour pressures (high p/p_0 values) multimolecular adsorption takes place. Since the enthalpy of adsorption is close to the enthalpy of condensation, the properties of the adsorbate are similar to those of a liquid.

To understand the conditions under which condensation of adsorbate vapour in the form of a liquid film may take place in the capillaries we must analyse the dependence of the vapour pressure on the curvature of the liquid surface. Figure 3.17 shows the forces acting on a liquid molecule lying on concave, flat and convex liquid surfaces. We see that the sphere of forces

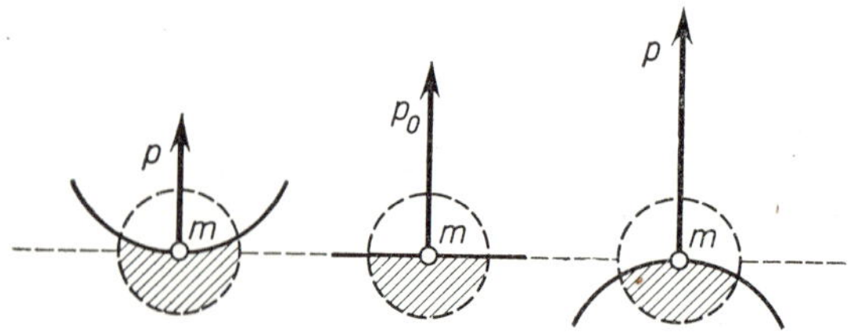

Fig. 3.17 Diagram showing the forces acting on a liquid molecule on a concave, flat and convex surface, respectively.

of molecular interaction of that molecule lies chiefly in the liquid and gas phases in the cases of concave and convex surfaces, respectively. We conclude that liquid molecules lying on a concave surface are held more strongly by the liquid phase than when the surface is flat. We observe the opposite effect when the liquid surface is convex. Therefore the saturated vapour pressure over a concave surface is smaller, and over a convex surface greater, as compared with a flat surface. Small drops of a liquid always show greater vapour pressures than large liquid surfaces, which is why the former evaporate more rapidly.

Adsorption layers covering the capillary walls of a porous adsorbent have a concave surface and the saturated vapour pressure over them is lower than over a flat surface. Therefore vapour condensation on these curved layers takes place at pressures lower than the p_0 of the adsorbed substance. The phenomenon of vapour condensation in capillaries at a pressure lower than the saturated vapour pressure p_0 which is reached over a flat surface is referred to as capillary condensation.

Thermodynamic considerations allow us to relate the saturated vapour pressures p and p_0 over a curved and a flat liquid surface, respectively. This relationship was studied by Kelvin [45] who derived equation:

$$\ln \frac{p}{p_0} = -\frac{2\gamma}{r_l} \frac{V_m}{RT} \quad (3.77)$$

where r_l is the radius of the liquid curvature (measured from the gas phase side; for a concave surface $r_l > 0$ and for a convex surface $r_l < 0$; for a flat liquid surface $r_l = \infty$), γ is the surface tension and V_m is the molar volume of the liquid. Relation (3.77) is known as the Kelvin equation.

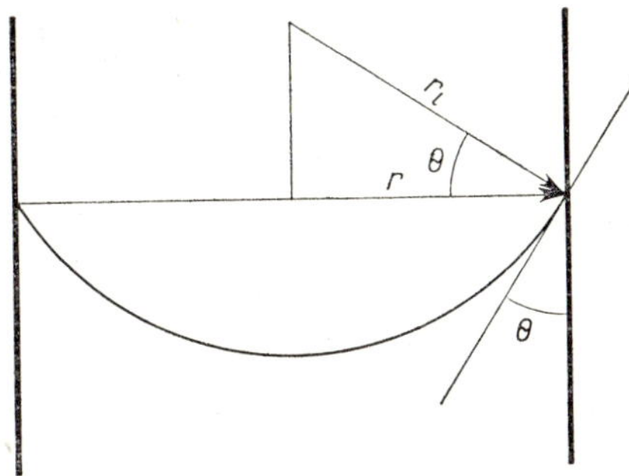

FIG. 3.18 Relation between the capillary radius r and the liquid meniscus curvature radius r_l.

For a liquid in a cylindrical capillary the radius r of the capillary is simply related to the radius r_l of the meniscus curvature (Fig. 3.18):

$$r = r_l \cos\theta \quad (3.78)$$

where θ is the wetting (limiting) angle. The dependence of saturated vapour pressure on the radius of the capillary is then given by:

$$\ln \frac{p}{p_0} = -\frac{2\gamma}{r} \frac{V_m}{RT} \cos\theta. \quad (3.79)$$

For a liquid which strongly wets capillary walls, $\theta = 0$, and hence $\cos\theta = 1$. The radius of the capillary is then practically equal to that of the curvature of the liquid meniscus.

Equation (3.77) has been derived for a curved liquid surface forming part of the surface of a sphere – the outer one in the case of a convex surface and the inner one for a concave surface. Hence

$$\ln \frac{p_{sph}}{p_0} = -\frac{2\gamma}{r_{sph}} \frac{V_m}{RT} \tag{3.80}$$

where p_{sph} and r_{sph} refer to the liquid curvature derived from the spherical surface.

If the liquid meniscus is cylindrical, then equation (3.77) has the form

$$\ln \frac{p_c}{p_0} = -\frac{\gamma}{r_c} \frac{V_m}{RT}. \tag{3.81}$$

From equation (3.81) (Cohan's equation [46]), it follows that the saturated vapour pressure above the liquid surface of cylindrical (concave) shape decreases less than that above a concave liquid surface which is a spherical sector, i.e. $p_c > p_{sph}$. This produces capillary condensation hysteresis (particularly reversible hysteresis), which is very common and is generally associated with the shape of the adsorbent capillaries.

Consider the simplest model of capillary condensation proposed by Cohan [46]. Let us assume that on the adsorbent surface there are conically shaped capillaries and cylindrical capillaries open at one and at both ends (Fig. 3.19).

As a result of adsorption of the vapour of a liquid which wets the adsorbent surface ($\theta = 0$), a curved (concave) adsorption layer is formed on the

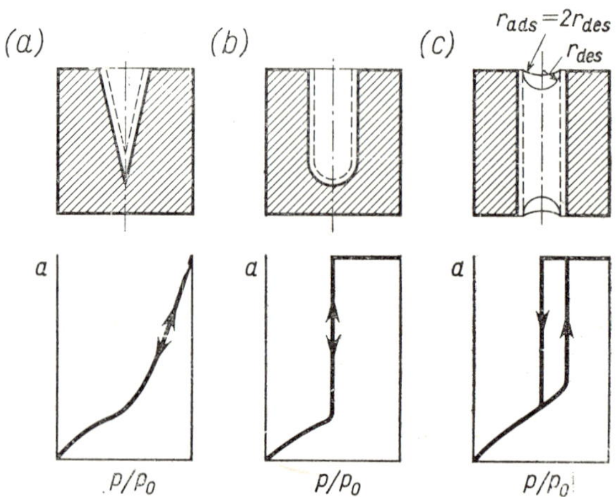

FIG. 3.19 Dependence of the shape of the adsorption–desorption isotherms on the porous structure of the adsorbent: r_{ads} are the radii of cylindrical menisci formed in the course of adsorption in capillaries with both ends open, r_{des} are the radii of spherical menisci formed in the process of desorption.

walls of the conical capillary. This surface has its greatest curvature at the narrowest end of the capillary, where a meniscus is formed whose shape derives from the spherical surface of radius r_{sph}. When the pressure of the adsorbate vapour over this surface attains a value consistent with equation (3.80), i.e. if:

$$\ln p_{sph} = \ln p_0 - \frac{2\gamma}{r_{sph}} \frac{V_m}{RT} \tag{3.82}$$

then condensation begins, leading to the displacement of the liquid meniscus to a wider part of the capillary, i.e. to the increase of r_{sph}. Thus, to allow further condensation of the vapour, its pressure must be increased (Fig. 3.19a). During desorption the process follows the same path, but in the opposite direction. Thus capillary condensation in conical (or tapered) pores is reversible.

As a result of adsorption in a cylindrical capillary closed at one end (resembling a test tube) a meniscus of shape derived from the spherical surface (as in a conical capillary) is formed at the closed end. When the vapour pressure attains the value p_{sph} given in equation (3.82), capillary condensation begins. The liquid in the capillary rises but the radius of the meniscus curvature remains unchanged. The whole capillary fills with liquid at constant pressure p_{sph}; the isotherm of capillary condensation is thus determined by a vertical line (Fig. 3.19b). Desorption follows the same path in the reverse direction.

If vapour adsorption proceeds in a cylindrical capillary open at both ends, then the shape of the meniscus formed is not part of spherical surface, and capillary condensation begins on the cylindrical meniscus of the adsorbate layer covering the capillary walls, at the pressure determined by the Cohan equation:

$$\ln p_c = \ln p_0 - \frac{\gamma}{r_c} \frac{V_m}{RT}. \tag{3.83}$$

In this case condensation leads to a thickening of the layer, i.e. decrease of the radius r_c. Therefore at pressure p_c the whole capillary fills with liquid. The capillary condensation isotherm occurring in the course of adsorption will, as before, be vertical. Since $p_c > p_{sph}$, this vertical line lies in the region of higher vapour pressure (Fig. 3.19c) than in the previous case. After filling the capillary, liquid menisci of spherical shape are formed at its ends, corresponding to the pressure $p_{sph} = p_c$, and accordingly $r_{sph} = 2r_c$. As the pressure increases from p_c to p_0 the curvature of these menisci will decrease with condensation of only a further small amount of vapour.

The desorption process proceeds initially in the reverse direction; on evaporation of small quantities of liquid, "spherical" menisci of increasing curvature form at both ends of the capillary. However, at $p_{sph}=p_c$ these menisci cannot yet disrupt, and therefore at such a vapour pressure the capillary still remains full. The desorption branch of the isotherm then departs from the adsorption branch, and the radius of the meniscus continues to decrease. Only when the vapour pressure decreases to p_{sph} as determined by equation (3.82), does the radius of the spherical meniscus, r_{sph}, become equal to the radius r_c, and all the liquid which has condensed in the capillary then evaporates. At vapour pressure $p_{sph}(p_{sph}<p_c)$ the desorption branch falls perpendicularly to the reversible polymolecular adsorption isotherm, and we obtain the characteristic capillary condensation hysteresis loop.

In real adsorbents, e.g. in adsorbents of a globular structure (Fig. 3.20), with a sufficient number of globule contact points, capillaries about these points are wedge-shaped. Therefore, in the vicinity of globule contact

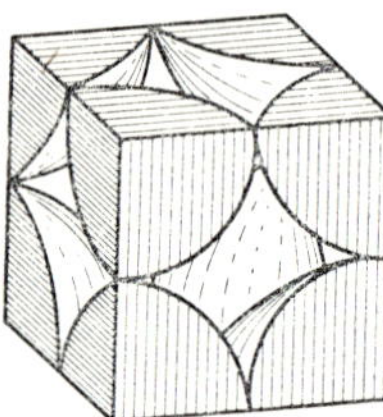

FIG. 3.20 Shape of pores in an adsorbent globular structure (the number of adjacent globules $n=6$).

points, the capillary condensation is reversible. However, at a certain distance from these points, as the capillaries become filled, menisci are formed of a near-cylindrical shape. The spaces between the globules are filled stepwise with liquid (as in a cylindrical capillary open at both ends) at a vapour pressure p_c close to that given by equation (3.83). In desorption, "spherical" menisci are formed at the outlets ending the capillaries. These menisci burst only when the vapour pressure decreases to the value p_{sph} given by equation (3.82), i.e. at $p_{sph}<p_c$. This leads to the formation of the capillary condensation hysteresis loop.

Adsorbents never have a completely uniform structure, and therefore their capillaries are not simultaneously filled. This is why the hysteresis loop has a slope as shown in Fig. 3.21, which demonstrates the adsorption–desorption isotherm for benzene vapour on uniform wide-porous silica gel.

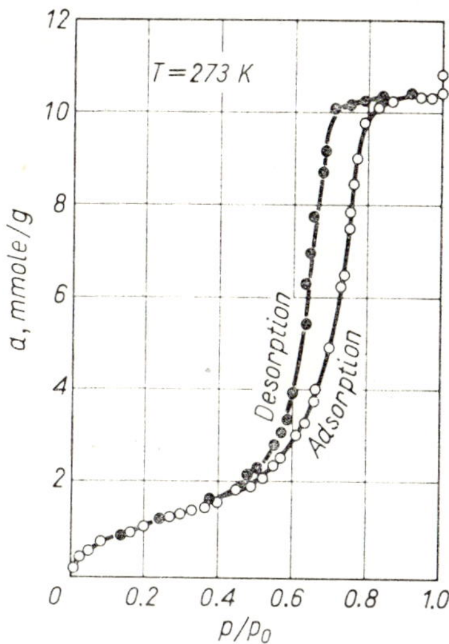

FIG. 3.21 Adsorption–desorption isotherm for benzene vapour and wide-porous silica gel (after Isirikyan and Kiselev [47]).

De Boer [48] has classified capillary hysteresis loops by their shapes, relating the latter to the occurrence of pores of given types. He distinguishes five types of hysteresis loops with at least one vertical or steep branch (the adsorption or desorption one) (Fig. 3.22). Most important are three types of loops, i.e. A, B and E.

According to de Boer, A-type hysteresis loops are obtained when the capillaries are regular or irregular cylinders or prisms open at both ends; B-type hysteresis loops appear on adsorption–desorption isotherms when the adsorbent pores are slit-shaped with parallel walls. Capillary condensation hysteresis loops of E-type are obtained when the adsorbent pores are of ink-bottle shape or have the form of deformed tubes with narrow ends and other contractions. The remaining two types of hysteresis loops (C and D) correspond to pores of shapes as for hysteresis loops A or B but partly deformed, e.g. the walls of slit-type pores are not parallel, etc.

The experimental hysteresis loops of adsorption–desorption isotherms are thus a combination of two or more types as distinguished by de Boer in terms of the non-homogeneous porous structure of the adsorbents.

Isotherms of gas and vapour adsorption on zeolites attract interest because these adsorbents act as molecular sieves, and therefore adsorption of molecules of different sizes and shapes proceeds in different ways. Figure 3.23 presents adsorption isotherms for benzene and *n*-hexane vapours

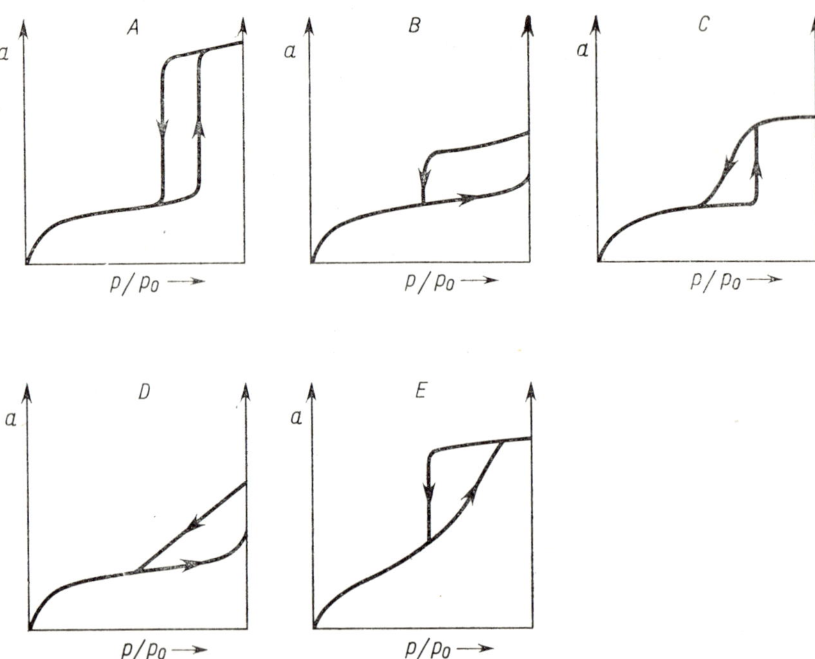

FIG. 3.22 De Boer's classification of capillary condensation hysteresis loops (after de Boer [48]).

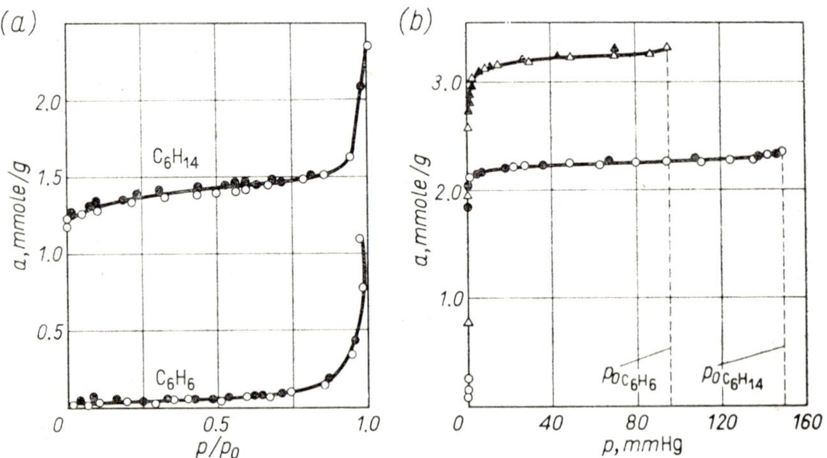

FIG. 3.23 Adsorption-desorption isotherms of benzene and *n*-hexane vapours on Zeolite 5A (a) and Zeolite 13X (b) (after Kiselev and Pavlova [49], and Zhdanov *et al.* [50]).

on Zeolite 5A and 13X. It appears that *n*-hexane molecules penetrate the internal spaces of both types of zeolites, and therefore their vapour adsorption isotherms differ only slightly. Microporous Zeolite 5A does not adsorb benzene vapour because benzene molecules are too large to pass through the outlets of the zeolite capillaries. In contrast, the wide spaces of Zeolite 13X can be penetrated by benzene molecules, and its vapour adsorption isotherm lies above that of *n*-hexane, because benzene molecules show a stronger interaction with the zeolite surface.

The capillaries of microporous adsorbents become filled for small relative vapour pressures p/p_0 and adsorption rapidly reaches a maximum. This is distinctly seen in the case of porous zeolite crystals.

3.5 Thermodynamics of Gas and Vapour Adsorption

3.5.1 Introductory Considerations

The application of thermodynamic methods to gas and vapour physical adsorption on solid adsorbents began only in 1940. Contemporary developments in the thermodynamics of adsorption processes, and in particular of physical adsorption, are due primarily to Kiselev [51], Hill [52, 53], Everett [54, 55] and Schay [56].

Before thermodynamics can be applied to adsorption processes, the adsorption system must be suitably selected and defined. One can, for instance, choose an adsorption chamber with a known quantity of adsorbent and a given quantity of gas or vapour at a given equilibrium pressure. In such a system, one does not distinguish between adsorbed and non-adsorbed gas. Although the thermodynamic description is complete, it is of little use for the molecular interpretation of the phenomenon.

A more useful approach to the problem assumes that the transition from the non-adsorbed gas to the surface layer proceeds in a stepwise fashion.

In adsorption, the simplest thermodynamic system consists of adsorbed gas constituting a separate phase in equilibrium with non-adsorbed gas (bulk phase) [57, 58]. Let us then consider the thermodynamics of adsorption involving a single gas, although adsorption of two-component mixtures is also discussed in the literature [52, 59, 60].

If we describe the state of n^s moles of adsorbed gas in terms of temperature T, gas pressure in the volume phase p (external pressure), entropy S^s, volume V^s, adsorbent surface area A, and interfacial tension γ (or surface pressure π), then we have eight characteristic thermodynamic functions (thermodynamic potentials) [61, 62].

3.5.2 Basic Thermodynamic Relations

Initially let us assume that the thermodynamic system consists of n^s moles of adsorbed gas (in the surface layer) in equilibrium with the bulk phase gas at pressure p. We also assume that the adsorbent is neutral and that its surface area A is proportional to its amount (or number of moles n_a). Changes of surface area $\mathrm{d}A$ mean that an appropriate amount $\mathrm{d}n_a$ of adsorbent is introduced into or removed from the system in an unchanged form. We can then derive the following equation for the internal energy change:

$$\mathrm{d}U^s = T\,\mathrm{d}S^s - p\,\mathrm{d}V^s + \gamma\,\mathrm{d}A + \mu^s\,\mathrm{d}n^s. \tag{3.84}$$

According to the definition of Helmholtz free energy $F^s = U^s - TS^s$,

$$\mathrm{d}F^s = \mathrm{d}U^s - \mathrm{d}(TS^s) = \mathrm{d}U^s - T\,\mathrm{d}S^s - S^s\,\mathrm{d}T$$

or

$$\mathrm{d}F^s = -S^s\,\mathrm{d}T - p\,\mathrm{d}V^s + \gamma\,\mathrm{d}A + \mu^s\,\mathrm{d}n^s. \tag{3.85}$$

The internal energy U^s of adsorbed gas (surface layer) is therefore the characteristic function at constant values of S^s, V^s, A and n^s, whereas the Helmholtz free energy F^s is its characteristic function at constant values of T, V^s, A and n^s, since under such conditions, at thermodynamic equilibrium we have $\mathrm{d}U^s = 0$ and $\mathrm{d}F^s = 0$, respectively.

However, as distinct from the bulk phase, the properties of the surface layer (or, generally, the interfacial layer) depend on two rather than one pair of mechanical variables p and V^s, and γ and A. In each pair, one variable is intensive (p or γ) and one extensive (V^s or A). Therefore three definitions of enthalpy and three definitions of the Gibbs free energy of the surface layer are possible [62], viz.

(i)

$$\mathscr{H}^s = U^s + pV^s, \tag{3.86}$$

hence

$$\mathrm{d}\mathscr{H}^s = \mathrm{d}U^s + \mathrm{d}(pV^s) = \mathrm{d}U^s + p\,\mathrm{d}V^s + V^s\,\mathrm{d}p$$

or

$$\mathrm{d}\mathscr{H}^{\mathrm{s}}=T\,\mathrm{d}S^{\mathrm{s}}+V^{\mathrm{s}}\,\mathrm{d}p+\gamma\,\mathrm{d}A+\mu^{\mathrm{s}}\,\mathrm{d}n^{\mathrm{s}}. \tag{3.87}$$

The enthalpy $\mathscr{H}^{\mathrm{s}}$ is thus a characteristic fuction of the surface layer when S^{s}, p, A and n^{s} are constant.

(ii)

$$\hat{H}^{\mathrm{s}}=U^{\mathrm{s}}-\gamma A, \tag{3.88}$$

hence

$$\mathrm{d}\hat{H}^{\mathrm{s}}=\mathrm{d}U^{\mathrm{s}}-\mathrm{d}(\gamma A)=\mathrm{d}U^{\mathrm{s}}-\gamma\,\mathrm{d}A-A\,\mathrm{d}\gamma$$

or

$$\mathrm{d}\hat{H}^{\mathrm{s}}=T\,\mathrm{d}S^{\mathrm{s}}-p\,\mathrm{d}V^{\mathrm{s}}-A\,\mathrm{d}\gamma+\mu^{\mathrm{s}}\,\mathrm{d}n^{\mathrm{s}}. \tag{3.89}$$

The enthalpy $\hat{H}^{\mathrm{s}}$ is the characteristic function of the surface layer when S^{s}, γ, V^{s} and n^{s} are constant.

(iii)

$$H^{\mathrm{s}}=U^{\mathrm{s}}+\mathrm{p}V^{\mathrm{s}}-\gamma A \tag{3.90}$$

or

$$\mathrm{d}H^{\mathrm{s}}=T\,\mathrm{d}S^{\mathrm{s}}+V^{\mathrm{s}}\,\mathrm{d}p-A\,\mathrm{d}\gamma+\mu^{\mathrm{s}}\,\mathrm{d}n^{\mathrm{s}}. \tag{3.91}$$

Enthalpy H^{s} defined in this way is the characteristic function of the surface layer when S^{s}, p, γ and n^{s} are constant.

The three definitions of Gibbs free energy are given similarly,

$$\mathscr{G}^{\mathrm{s}}=F^{\mathrm{s}}+\mathrm{p}V^{\mathrm{s}}, \tag{3.92}$$

$$\hat{G}^{\mathrm{s}}=F^{\mathrm{s}}-\gamma A, \tag{3.93}$$

$$G^{\mathrm{s}}=F^{\mathrm{s}}+pV^{\mathrm{s}}-\gamma A. \tag{3.94}$$

The corresponding differentials are:

$$\mathrm{d}\mathscr{G}^{\mathrm{s}}=-S^{\mathrm{s}}\,\mathrm{d}T+V^{\mathrm{s}}\,\mathrm{d}p+\gamma\,\mathrm{d}A+\mu^{\mathrm{s}}\,\mathrm{d}n^{\mathrm{s}}, \tag{3.95}$$

$$\mathrm{d}\hat{G}^{\mathrm{s}}=-S^{\mathrm{s}}\,\mathrm{d}T-p\,\mathrm{d}V^{\mathrm{s}}-A\,\mathrm{d}\gamma+\mu^{\mathrm{s}}\,\mathrm{d}n^{\mathrm{s}}, \tag{3.96}$$

$$\mathrm{d}G=-S^{\mathrm{s}}\,\mathrm{d}T+V^{\mathrm{s}}\,\mathrm{d}p-A\,\mathrm{d}\gamma+\mu^{\mathrm{s}}\,\mathrm{d}n^{\mathrm{s}}. \tag{3.97}$$

Hence $\mathscr{G}^{\mathrm{s}}$ is a characteristic function of the surface layer for constant T, p, A and n^{s}; $\hat{G}^{\mathrm{s}}$ that for constant T, γ, V^{s} and n^{s}; and G^{s} that for constant T, p, γ and n^{s}. In the case of the surface (interfacial) layer we have, in contrast to the bulk phases, eight, not four, characteristic thermodynamic functions. The functions $\hat{H}^{\mathrm{s}}$ and $\hat{G}^{\mathrm{s}}$ are, however, seldom used.

Let us define the Gibbs free energy of the surface layer by equation (3.92). Denoting the molar differential quantities as:

$$\bar{S}^{\mathrm{s}}_{\mathrm{m}}=\left(\frac{\partial S^{\mathrm{s}}}{\partial n^{\mathrm{s}}}\right)_{p,T,A}, \quad \bar{V}^{\mathrm{s}}_{\mathrm{m}}=\left(\frac{\partial V^{\mathrm{s}}}{\partial n^{\mathrm{s}}}\right)_{p,T,A} \quad \text{and} \quad \mu^{\mathrm{s}}=\bar{G}^{\mathrm{s}}_{\mathrm{m}}=\left(\frac{\partial \mathscr{G}^{\mathrm{s}}}{\partial n^{\mathrm{s}}}\right)_{p,T,A}$$

and differentiating equation (3.95) with respect to n^s, we get:

$$d\mu^s = -\bar{S}^s_m\, dT + \bar{V}^s_m\, dp - \left(\frac{\partial \gamma}{\partial n^s}\right)_{p,T,A} dA + \left(\frac{\partial \mu^s}{\partial n^s}\right)_{p,T,A} dn^s. \qquad (3.98)$$

Consider variation of p and T for a constant amount of gas ($dn^s=0$) adsorbed on a fixed amount of adsorbent of given surface area ($dA=0$). Then:

$$d\mu^s = -\bar{S}^s_m\, dT + \bar{V}^s_m\, dp. \qquad (3.99)$$

Let us now extend our thermodynamic system involving the surface layer (adsorbed gas) to include an equilibrium with the bulk (gas) phase. Then $d\mu^s = d\mu$, where μ is the chemical potential of non-adsorbed gas, i.e.:

$$-\bar{S}^s_m\, dT + \bar{V}^s_m\, dp = -S_m\, dT + V_m\, dp \qquad (3.100)$$

where $S_m = S/n$ is the molar entropy of the gas in the bulk phase, $V_m = V/n$ is the molar volume of the gas in that phase, and n is the number of moles of non-adsorbed gas. Usually $V_m \gg V^s_m$, so, if we assume that $pV_m = RT$, then:

$$\Delta_a S^s_{m,A} = \bar{S}^s_m - S_m = -RT\left(\frac{\partial \ln p}{\partial T}\right)_{n^s,A} \qquad (3.101)$$

where $\Delta_a S^s_{m,A}$ is called the differential molar entropy of gas adsorption, expressing the entropy change due to the virtual transfer to the surface layer of one mole of gas in equilibrium with the bulk phase, the surface concentration of adsorbed gas remaining constant (hence n^s=const, and A=const). Such an entropy change can be determined from the adsorption isosteres.

The Gibbs–Duhem equation for the surface layer, relating variations in intensive quantities, can be written as [61]:

$$S^s\, dT - V^s\, dp + A\, d\gamma + n^s\, d\mu^s = 0. \qquad (3.102)$$

After dividing equation (3.102) by n^s and introducing $S^s/n^s = S^s_m$, $V^s/n^s = V^s_m$ and $n^s/A = \alpha^s$ we get:

$$S^s_m\, dT - V^s_m\, dp + \frac{1}{\alpha^s}\, d\gamma + d\mu^s = 0. \qquad (3.103)$$

At thermodynamic equilibrium:

$$d\mu^s = d\mu = -S_m\, dT + V_m\, dp.$$

Substituting the above expression for $d\mu^s$ into equation (3.103) we get:

$$S^s_m dT - V^s_m dp + \frac{1}{\alpha^s} d\gamma - S_m dT + V_m dp = 0 \tag{3.104}$$

or

$$(S^s_m - S_m) dT + (V_m - V^s_m) dp + \frac{1}{\alpha^s} d\gamma = 0. \tag{3.105}$$

Two important conclusions follow from equation (3.105). Firstly, for $T = \text{const}$ this equation becomes:

$$d(\gamma)_T = -\alpha^s (V_m - V^s_m) d(p)_T \tag{3.106}$$

and represents the Gibbs adsorption isotherm (cf. Chapter 2) for the solid/gas interface. We may determine γ from the adsorption isotherms ($\alpha^s = f(p)_T$), and if we assume $V_m \gg V^s_m$ and $pV_m = RT$, then:

$$\gamma = -RT \int_0^p \alpha^s d\ln p. \tag{3.107}$$

A second important conclusion can be drawn from equation (3.105) when $d\gamma = 0$. We find:

$$\Delta_a S^s_{m,\gamma} = S^s_m - S_m = -RT \left(\frac{\partial \ln p}{\partial T}\right)_\gamma \tag{3.108}$$

where $\Delta_a S^s_{m,\gamma}$ is the differential molar entropy of gas adsorption at constant interfacial tension γ, S^s_m is the molar entropy of adsorbed gas; $\Delta_a S_{m,\gamma}$ is the entropy change corresponding to virtual transfer (at equilibrium) of one mole of gas from the gas phase to the surface layer at $\gamma = \text{const}$. Its value may be determined from a suitable set of adsorption isotherm by means of equation (3.107).

Consider now the transfer of internal energy to the environment related to gas adsorption, i.e. the heat of adsorption. As in the case of entropy changes, we consider the case of thermodynamic equilibrium. We thus consider heat effects related to the virtual transfer of one mole (at equilibrium) from the volume phase to the surface layer, some variables remaining constant. The differential molar quantities involved depend on the process parameters.

If in the isothermal and isochoric adsorption of dn^s moles of gas the heat involved is $d\Delta_a U^s$ (where $\Delta_a U^s = U^s - U$, and U is the internal energy of the non-adsorbed gas), then the differential molar internal energy of gas adsorption can be defined as

$$\Delta_a U_m^s = \left(\frac{\partial \Delta_a U^s}{\partial n^s}\right)_{T, A, V^s} . \tag{3.109}$$

Let us assume a fixed quantity of adsorbent, i.e. A=const. Then if the gas in the bulk phase is ideal ($U = nU_m$) and noting that $dn = -dn^s$, we can write:

$$\Delta_a U_m^s = \left(\frac{\partial U^s}{\partial n^s}\right)_{T, A, V^s} - U_m = \bar{U}_m^s - U_m \tag{3.110}$$

where $\bar{U}_m^s$ is the differential molar internal energy of adsorbed gas and U_m is the molar internal energy of gas in the bulk phase. $\Delta_a U_m^s$ is often referred to as differential heat of adsorption and denoted by q_d (with $\Delta_a U_m^s = -q_d$) [52]. Hence, $\Delta_a U_m^s$ is the heat involved in the transfer of 1 mole of gas from the bulk phase to the interface at thermodynamic equilibrium (virtual transfer), if V, V^s and adsorbed surface area A are constant.

Let us introduce the concept of the "surface piston" by means of which dn^s moles of gas are transferred in adsorption from the bulk phase to the surface layer at γ=const. If the equilibrium pressure p and temperature T are constant and noting that $dn = -dn^s$, then the virtual transfer of 1 mole of gas from the bulk phase to the surface layer is accompanied by an enthalpy change:

$$\Delta_a H_m^s = \frac{H^s}{n^s} - \frac{H}{n} = H_m^s - H_m = T\Delta_a S_{m,\gamma}^s \tag{3.111}$$

where the enthalpy H^s is defined by equation (3.90). $\Delta_a H_m^s$ is often referred to as the equilibrium enthalpy (heat) of adsorption [53, 54]. By comparing equations (3.108) and (3.111) we obtain an expression for the determination of $\Delta_a H_m^s$ from a set of adsorption isotherms measured at various temperatures:

$$\Delta_a H_m^s = -RT^2 \left(\frac{\partial \ln p}{\partial T}\right)_\gamma = R \left(\frac{\partial \ln p}{\partial \dfrac{1}{T}}\right)_\gamma . \tag{3.112}$$

If the adsorption system is provided with a "piston", so that the isothermal process in simultaneously isobaric, then we can define the differential molar enthalpy of adsorption as:

$$\Delta_a \mathscr{H}_m^s = \left(\frac{\partial \Delta \mathscr{H}^s}{\partial n^s}\right)_{T, A, p} \tag{3.113}$$

where $\mathrm{d}\Delta\mathscr{H}^{\mathrm{s}}$ ($\Delta\mathscr{H}^{\mathrm{s}}=\mathscr{H}^{\mathrm{s}}-H$, H being the enthalpy of the gas in the bulk phase) is the enthalpy involved in the adsorption of $\mathrm{d}n^{\mathrm{s}}$ moles of gas under the given conditions and at $A=\mathrm{const}$. Assuming as before that the gas in the bulk phase is ideal, (i.e. $H=nH_{\mathrm{m}}$) and noting that $\mathrm{d}n=-\mathrm{d}n^{\mathrm{s}}$, we have

$$\Delta_{\mathrm{a}}\mathscr{H}^{\mathrm{s}}_{\mathrm{m}}=\left(\frac{\partial\mathscr{H}^{\mathrm{s}}}{\partial n^{\mathrm{s}}}\right)_{T,A,p}-H_{\mathrm{m}}=\bar{H}^{\mathrm{s}}_{\mathrm{m}}-H_{\mathrm{m}}. \tag{3.114}$$

where $\Delta_{\mathrm{a}}\mathscr{H}^{\mathrm{s}}_{\mathrm{m}}$ is therefore the enthalpy involved in the virtual transfer (at equilibrium) of 1 mole of gas from the bulk phase to the surface layer at constant temperature, pressure and adsorbent surface area. At equilibrium,

$$\Delta_{\mathrm{a}}\mathscr{H}^{\mathrm{s}}_{\mathrm{m}}=T\Delta_{\mathrm{a}}S^{\mathrm{s}}_{\mathrm{m},A}. \tag{3.115}$$

Comparing this equation with equation (3.101), we conclude that when the gas phase is ideal:

$$\Delta_{\mathrm{a}}\mathscr{H}^{\mathrm{s}}_{\mathrm{m}}=-RT^2\left(\frac{\partial\ln p}{\partial T}\right)_{\alpha^{\mathrm{s}}}=RT\left(\frac{\partial\ln p}{\partial\frac{1}{T}}\right)_{\alpha^{\mathrm{s}}}. \tag{3.116}$$

Thus $\Delta_{\mathrm{a}}\mathscr{H}^{\mathrm{s}}_{\mathrm{m}}$ can be determined from adsorption isosteres (like $\Delta_{\mathrm{a}}S^{\mathrm{s}}_{\mathrm{m},A}$), and this quantity is called the isosteric heat of adsorption, denoted by q_{st}, where $\Delta_{\mathrm{a}}\mathscr{H}^{\mathrm{s}}_{\mathrm{m}}=-q_{\mathrm{st}}$ [52, 53]. Equation (3.116), like equation (3.112) is a version of the Clausius–Clapeyron equation describing thermodynamic phase changes (state changes).

From the fundamental relationship between the internal energy and enthalpy it follows that:

$$\Delta_{\mathrm{a}}\mathscr{H}^{\mathrm{s}}_{\mathrm{m}}=(\bar{U}^{\mathrm{s}}_{\mathrm{m}}+p\bar{V}^{\mathrm{s}}_{\mathrm{m}})-(U_{\mathrm{m}}-pV_{\mathrm{m}})=\Delta_{\mathrm{a}}U^{\mathrm{s}}_{\mathrm{m}}+p(\bar{V}^{\mathrm{s}}_{\mathrm{m}}-V_{\mathrm{m}}).$$

Assuming that $V_{\mathrm{m}}\gg\bar{V}^{\mathrm{s}}_{\mathrm{m}}$ and $pV_{\mathrm{m}}=RT$ we finally get:

$$\Delta_{\mathrm{a}}\mathscr{H}^{\mathrm{s}}_{\mathrm{m}}=\Delta_{\mathrm{a}}U^{\mathrm{s}}_{\mathrm{m}}-RT. \tag{3.117}$$

The relation between $\Delta_{\mathrm{a}}H^{\mathrm{s}}_{\mathrm{m}}$ and $\Delta_{\mathrm{a}}\mathscr{H}^{\mathrm{s}}_{\mathrm{m}}$ may easily be found by adding to and subtracting from equation (3.111) the product $T\bar{S}^{\mathrm{s}}_{\mathrm{m}}$:

$$\Delta_{\mathrm{a}}H^{\mathrm{s}}_{\mathrm{m}}=TS^{\mathrm{s}}_{\mathrm{m}}-TS_{\mathrm{m}}+T\bar{S}^{\mathrm{s}}_{\mathrm{m}}-T\bar{S}^{\mathrm{s}}_{\mathrm{m}} \tag{3.118}$$

or

$$\Delta_{\mathrm{a}}H^{\mathrm{s}}_{\mathrm{m}}=T(\bar{S}^{\mathrm{s}}_{\mathrm{m}}-S_{\mathrm{m}})+T(S^{\mathrm{s}}_{\mathrm{m}}-\bar{S}^{\mathrm{s}}_{\mathrm{m}}). \tag{3.119}$$

Using equation (3.115) we get:

$$\Delta_{\mathrm{a}}H^{\mathrm{s}}_{\mathrm{m}}=\Delta_{\mathrm{a}}\mathscr{H}^{\mathrm{s}}_{\mathrm{m}}+T(S^{\mathrm{s}}_{\mathrm{m}}-\bar{S}^{\mathrm{s}}_{\mathrm{m}}). \tag{3.120}$$

Comparing the expressions for $d\mu^s$ as given by equations (3.99) and (3.103), yields a relationship between S^s_m and $\bar{S}^s_m$ (at constant p, A and n^s).

When $V^s_m = \bar{V}^s_m$, we get:

$$S^s_m - \bar{S}^s_m = -\frac{1}{\alpha^s}\left(\frac{\partial \gamma}{\partial T}\right)_{p,\,\alpha^s}. \tag{3.121}$$

Since the difference $S^s_m - \bar{S}^s_m$ is small, and in certain cases close to zero, both $\Delta_a H^s_m$ and $\Delta_a \mathscr{H}^s_m$ refer in principle to the same (isosteric) heat of adsorption.

The heats of gas adsorption described above are isothermic. Kington and Aston [63] have defined the adiabatic differential heat of gas adsorption by:

$$q_a = (C_c + nC_{pm} + n^s C^s_{pm})\left(\frac{\partial T}{\partial n^s}\right)_{\text{adiab}} \tag{3.122}$$

in which C_c is the heat capacity of the calorimeter and adsorbent under constant pressure, C_{pm} and C^s_{pm} are the molar heat capacities of the gas in the bulk and surface phases. Then q_a and q_{st} are related by:

$$q_a = q_{st} + V\left(\frac{\partial p}{\partial n^s}\right)_{\text{adiab}}. \tag{3.123}$$

NOTE. Simple thermodynamic considerations allow us to relate the constant k in the Langmuir adsorption isotherm (or constant C in the BET equation) to the relevant energies of adsorption. For every equilibrium,

$$\Delta G = \Delta H - T\Delta S = -RT \ln K$$

where K is the equilibrium constant.

For adsorption equilibrium using the Langmuir theory ($\Delta_a H^s_m = H^s_m - H_m$, $\Delta_a S^s_m = S^s_m - S_m$) we get:

$$\Delta_a G^s_m = \Delta_a H^s_m - T\Delta_a S^s_m = -RT \ln k. \tag{3.124}$$

Hence

$$k = \exp\frac{-\Delta_a G^s_m}{RT} = \exp\frac{\Delta_a S^s_m}{R}\exp\frac{-\Delta_a H^s_m}{RT}. \tag{3.125}$$

Putting $g_0 = \exp \Delta_a S^s_m / R$ (the so-called entropic factor) we can write:

$$k = g_0 \exp\frac{-\Delta_a H^s_m}{RT}. \tag{3.126}$$

From BET theory $C = k'/k_L$, where k' is the adsorption equilibrium constant for the first layer of adsorbate, and k_L is the condensation constant. So

$$k' = g'_0 \exp\frac{-\Delta_a H^s_m}{RT} \tag{3.127}$$

and

$$k_L = g_{0L} \exp \frac{-\Delta H_m^L}{RT} \tag{3.128}$$

so if $g_0 = g'_0/g_{0L}$, then

$$C = g_0 \exp \frac{-(\Delta_a H_m^s - \Delta H_m^L)}{RT}. \tag{3.129}$$

3.5.3 Experimental Investigation of the Thermodynamic Properties of Adsorption Layers

3.5.3.1 *Thermodynamic Functions Determined from Adsorption Isotherms*

The differential molar enthalpy for gas or vapour adsorption (isosteric heat of adsorption) can be calculated from adsorption isotherms obtained for at least three different temperatures. From such isotherms, the dependence of log p vs. $1/T$ can be plotted (Fig. 3.24) for given values of α or given coverage values θ (for constant amount of adsorbent, i.e. A = const).

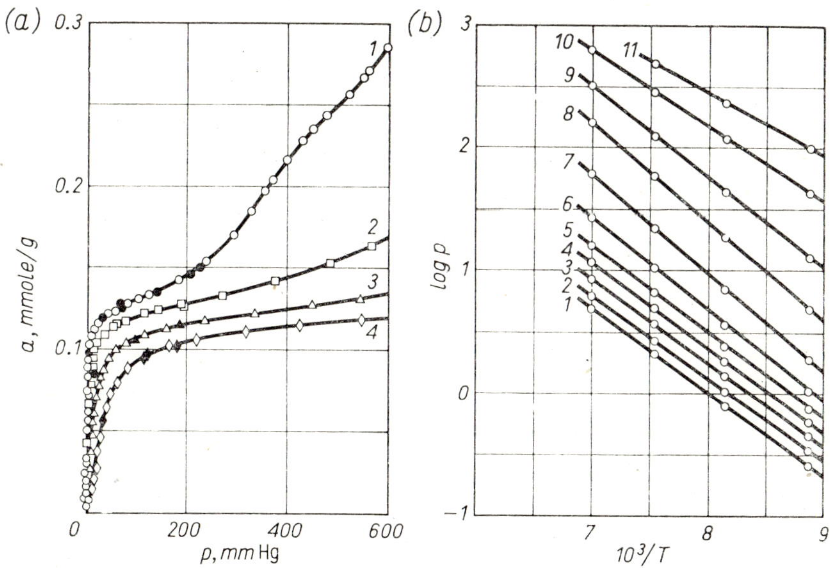

FIG. 3.24 (a) Methane adsorption isotherms on graphitized carbon black at: *1* – 113 K, *2* – 123 K, *3* – 133 K and *4* – 143 K. The full points refer to desorption (after Kiselev [33]). (b) Methane adsorption isosteres on graphitized carbon black for adsorption in mmole/g: *1* – 0.0075, *2* – 0.010, *3* – 0.015, *4* – 0.020, *5* – 0.030, *6* – 0.050, *7* – 0.080, *8* – 0.100, *9* – 0.110, *10* – 0.120, *11* – 0.130.

It turns out that, over a wide range of pressures and temperatures, the gas (vapour) adsorption isosteres plotted for such a co-ordinate system yield straight lines [64]. In accordance with equation (3.116), written in the form:

$$\Delta_a \mathscr{H}^s_m = 2.303R \left(\frac{\partial \log p}{\partial (1/T)}\right)_{\alpha^s} \tag{3.130}$$

the slope (tangent of inclination angle) of the straight line is proportional to $\Delta_a \mathscr{H}^s_m$. The differential molar entropy of adsorption is obtained from equation (3.115). In numerical calculations, equation (3.116) may be used in the form:

$$\frac{\Delta_a \mathscr{H}^s_m}{2.303R} = \frac{\Delta (\log p)}{\Delta (1/T)} \,. \tag{3.131}$$

We can also determine similarly the isothermal differential enthalpy of gas adsorption $\Delta_a H^s_m$ (the equilibrium heat of adsorption) and molar entropy of gas adsorption $\Delta_a S^s_{m,\gamma}$ using equation (3.112). To determine the value of γ equation (3.106) or (3.107) should be used. A knowledge of the absolute value of γ is not necessary, and it suffices to know the pressure at two temperatures at constant γ. We write equation (3.107) in the form:

$$\gamma = -K_0 RT \int_0^p V^s \mathrm{d}(\ln p)_T \tag{3.132}$$

where V^s denotes the volume of gas adsorbed on 1 g of adsorbent and K_0 is chosen so that $\alpha^s = V^s K_0$. The value of $\Delta_a H^s_m$ is determined as follows. Using equation (3.107) or (3.132) $\gamma/K_0 R$ (or γ) is plotted vs. p at various temperatures. From these curves, and using equation (3.112), one can determine $\Delta_a H^s_m$ by introducing the values of log p as a function of $1/T$ at constant γ. In numerical calculations, we can represent equation (3.112) analogously to equation (3.131), i.e.

$$\frac{\Delta_a H^s_m}{2.303R} = \frac{\Delta (\log p)_\gamma}{\Delta (1/T)_\gamma} \,. \tag{3.133}$$

Figure 3.25 represents the frequently used plot of isosteric heat of gas adsorption q_{st} against adsorbent coverage θ (or adsorbtion α^s). Experimental data indicate that the differential heat of adsorption of gases (vapours) on graphitized carbon black at small coverage decreases initially (the region of surface heterogeneity) and subsequently increases, reaching its maximum at $\alpha^s = \alpha^s_m$, corresponding to the filling of the first monolayer. With further increase in adsorption, q_{st} decreases quite rapidly until it reaches that value corresponding to the differential heat of condensation L.

The bends on the decreasing curves may point to the formation of subsequent adsorbate layers on the adsorbent surface.

The increase of q_{st} in the course of formation of the first monolayer may also be due to horizontal interaction between the molecules of the adsorbed gas (see Section 3.4.6) [33].

Capillary condensation is distinctly visible on the q_{st} vs. α^s plots (Fig. 3.26).

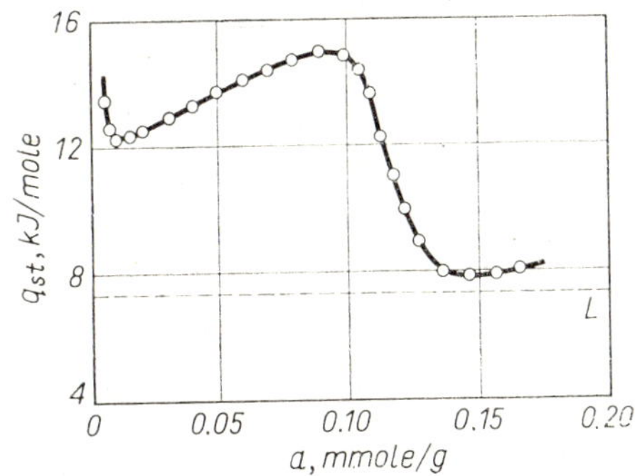

FIG. 3.25 Variation of the isosteric molar differential heat of adsorption (q_{st}) of methane on graphitized carbon black with adsorption α^s; $L = -\Delta H_m^L$ is the molar heat of condensation (after Jura and Harkins [38]).

The dependence of differential molar entropy of gas (vapour) adsorption on coverage θ (or α^s) represented in Fig. 3.27 is often a mirror image of the q_{st} vs. θ relation (note that $\Delta_a \mathcal{H}_m^s = -q_{st}$). Prior to adsorption and in the early stages of adsorption, the gas (vapour) molecules reveal a high mobility, and their entropy is also large. In adsorption the absolute value of $\Delta_a S_m^s$ rapidly decreases and after passing through several extrema (generally two) it reaches zero, i.e. the condition of the adsorption layer tends to that of a liquid.

Thermodynamic functions characterizing adsorption can also be calculated from a single gas adsorption isotherm, provided we know the temperature coefficient of volume expansion of the adsorbate [66]. In accordance with the Polanyi potential adsorption theory the adsorption potential remains constant over a wide range of temperatures, i.e. $(\partial \varepsilon / \partial T)_{V^s} = 0$. This indicates that the characteristic adsorption curve is temperature independent and that every adsorption isotherm uniquely determines the shape of that curve. The temperature coefficient of adsorption may then be determined.

Calculation of thermodynamic adsorption functions from isotherms is currently very common, since gas adsorption chromatography provides

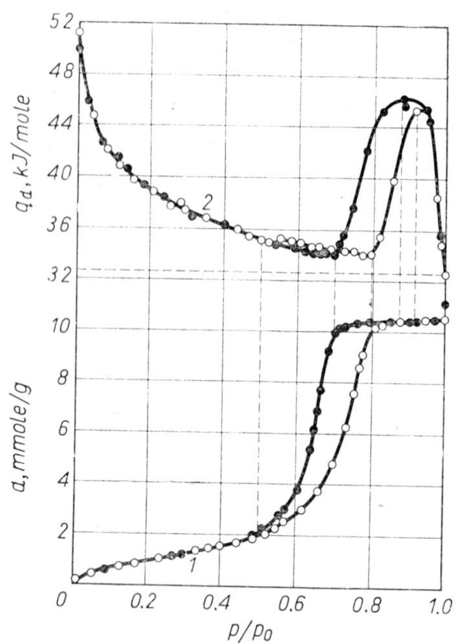

FIG. 3.26 Benzene vapour adsorption–desorption isotherm (*1*), and differential molar heat of adsorption (*2*) on wide-porous silica gel (after Isirikyan [65]).

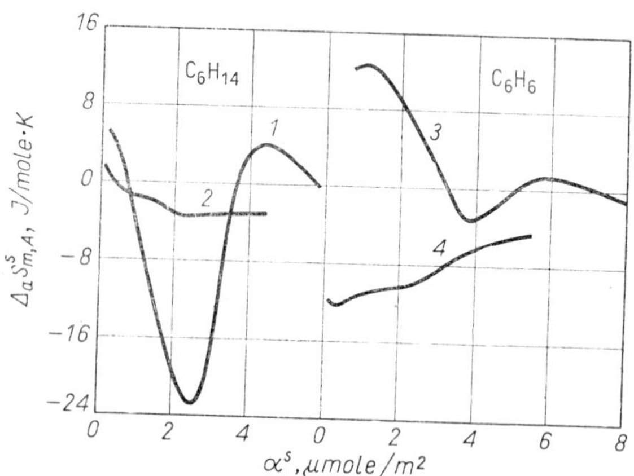

FIG. 3.27 Differential molar entropy of adsorption of *n*-hexane (curves *1* and *2*) and benzene (curves *3* and *4*) on graphitized carbon black (curves *1* and *3*) and silica gel (curves *2* and *4*) (after [15]).

for the rapid and accurate measurement of such isotherms. The differential molar heat of adsorption, for example, may be calculated directly from chromatographic data.

NOTE. Equations (3.101), (3.108), (3.112) and (3.116) and others give the differential molar thermodynamic adsorption functions for given temperature and pressure. Then: $\Delta_a \mathscr{G}^s_m = 0$ and $\Delta_a \mathscr{H}^s_m = T\Delta_a S^s_m$. The derived values of Gibbs free energy, enthalpy and entropy of gas adsorption are often represented in the form of differences of $\bar{G}^s_m$, $\bar{H}^s_m$ and $\bar{S}^s_m$ (or G^s_m, H^s_m and S^s_m) and the corresponding molar value in a certain reference (standard) state. In the case of vapour adsorption it is convenient to take as reference the given vapour under saturated vapour pressure p_0. If we assume that the vapour behaves ideally, then:

$$\Delta_a G^{s,\,p_0}_m = RT\ln h\,, \quad \Delta_a S^{s,\,p_0}_m = \Delta_a S^s_m - R\ln h \quad \text{and} \quad \Delta_a H^{s,\,p_0}_m = \Delta_a \mathscr{H}^s_m\,,$$

where $h = p/p_0$. Since the adsorbed vapour usually resembles a liquid more than a gas (when $p \rightarrow p_0$, the heat of adsorption tends to the heat of condensation), we can assume the liquid under the pressure of its saturated vapour p_0 as reference state. Then:

$$\Delta_a G^{s,\,L}_m = RT\ln h\,, \quad \Delta_a S^{s,\,L}_m = \Delta_a S^s_m - \Delta S^L_m - R\ln h \quad \text{and} \quad \Delta_a H^{s,\,L}_m = \Delta_a \mathscr{H}^s_m - \Delta H^L_m\,,$$

where $\Delta_a S^{s,\,L}_m$ and $\Delta_a H^{s,\,L}_m$ are the differential molar entropy and differential molar enthalpy, respectively, of so-called "pure adsorption", and ΔS^L_m and ΔH^L_m are the molar entropy and molar enthalpy of condensation of the adsorbate, respectively.

3.5.3.2 *Thermodynamic Functions Arising from Heats of Adsorption*

The results of calorimetric measurements of heats of adsorption may be used to calculate thermodynamic adsorption functions. This method has the advantage over the methods already described in that it allows for a more direct determination of the integral quantities and a much more accurate estimation of the heats of adsorption, and hence of the entropy. Denoting the integral heat of adsorption by ΔU^s we can calculate directly the differential molar heat of adsorption q_d in accordance with equation (3.122), as well as the remaining thermodynamic functions.

3.6 Adsorption from the Gaseous Phase on to Heterogeneous Adsorbent Surface

3.6.1 INTRODUCTION

Theories of adsorption from the gaseous phase usually assume homogeneity of the adsorbent surface, whereas, in reality, most adsorbent surfaces are energetically heterogeneous.

When discussing the energetical heterogeneity of adsorbent surfaces, the latter are generally classified into homogeneous and heterogeneous ones. When the adsorption energy (potential) $\varepsilon(x, y, z)$ is a function only of the distance z from the adsorbent surface, we say that the surface is homogeneous and uniform. If $\varepsilon(x, y, z)$ is, at a given distance z from the adsorbent surface, a periodic function of the vector of displacement parallel to the surface from point (x_0, y_0), corresponding to the active site (adsorption centre), then we say that the surface is homogeneous and non-uniform. In the case when the above mentioned function does not exist, we consider the surface as heterogeneous [67]. However, the terms uniformity and homogeneity as well as non-uniformity and heterogeneity of the adsorbent surface are often used as synonyms.

Langmuir [13] first drew attention to the energetical heterogeneity of adsorbent surfaces. He stated that his adsorption isotherm can be applied to any of a finite number of adsorbent surface sectors (patches) of equal energy per active site. The total adsorption isotherm may then be written:

$$a = \sum_i \frac{a_{m,i} k_i p}{1 + k_i p}. \tag{3.134}$$

When adsorption takes place on amorphous surfaces, the k_i vary continuously, and integration should be used instead of summation.

3.6.2 Models of the Heterogeneous Adsorbent Surface

After Langmuir many authors have published papers on how the effects of adsorbent surface energetical heterogeneity should be allowed for in adsorption isotherms. Studies on the energetic heterogeneity of adsorbent surface before 1960 have been reviewed by Young and Crowell [2]. The last twenty years have seen rapid developments in investigations on the effect of surface heterogeneity of solids on adsorption from the gaseous phase. These studies have concentrated on two basic models of surface heterogeneity [67, 68].

(i) *The Patchwise Topographical Distribution of Adsorption Centres* [69]

According to this model the adsorbent surface is composed of energetically homogeneous sectors (patches) within which adsorption conditions are identical. Every sector must have a surface area large enough to permit the neglect of interactions of molecules adsorbed at neighbouring homogeneous patches.

(ii) *The Random Topographical Distribution of Adsorption Centres* [67, 68]

Both heterogeneous surface models were compared with experimental data based on the virial formalism of gas adsorption. Fundamental studies in this area were carried out by Barker and Everett [70], Pierotti [71, 72], Sams [73], Steele [68] and Rudziński [74, 75], who revealed the great advantages of the virial description of gas adsorption on heterogeneous surfaces, allowing the investigation both of the total state of the surface and the distribution of adsorption centres over the surface. This requires, however, a knowledge of the temperature dependence of adsorption data [76 – 78].

3.6.3 Integral Adsorption Isotherm Equation

Most investigations of gas adsorption on heterogeneous adsorbent surface are based on the integral equation of the adsorption isotherm [79, 80] which is derived directly from equation (3.134):

$$\theta_t(p) = \int_{\Omega} \theta(p, \varepsilon)\, \chi(\varepsilon)\, d\varepsilon \tag{3.135}$$

where $\theta_t(p)$ is the total gas adsorption in terms of adsorbent surface coverage, $\theta(p, \varepsilon)$ is the gas adsorption isotherm for an arbitrarily chosen homogeneous patch of surface area of adsorption energy (potential) ε and is referred to as the local adsorption isotherm, $\chi(\varepsilon)$ is the adsorption energy distribution function characterizing the energetical heterogeneity of the adsorbent surface and Ω is the interval of possible adsorption energy changes. From physical considerations,

$$\int_{\Omega} \chi(\varepsilon)\, d\varepsilon = 1 \tag{3.136}$$

when the total adsorption is described by the function $\theta(p)$ and should equal the monolayer capacity if the total adsorption is determined by the quantity $a(p)$.

In equation (3.135) $\Omega = (0, \infty)$ is usually assumed as the integration interval [81, 82]. The energetical heterogeneity of the adsorbent surface estimated by this equation reduces to an analytic or numerical solution with respect to the $\chi(\varepsilon)$ function, having assumed the local adsorption model (adsorption in the homogeneous surface sector).

Local adsorption is generally described by the Langmuir equation in the form:

$$\theta(p,\varepsilon)=\left[1+\frac{K_H}{p}\exp\left(\frac{-\varepsilon}{RT}\right)\right]^{-1} \tag{3.137}$$

where K_H is the constant related to the distribution function of molecules for the adsorption monolayer. Taking account of equation (3.137) we get the adsorption isotherm integral equation in the form:

$$\theta(p)=\int_{\Omega}\left[1+\frac{K_H}{p}\exp\left(\frac{-\varepsilon}{RT}\right)\right]^{-1}\chi(\varepsilon)\,d\varepsilon. \tag{3.138}$$

There are two basic methods for solving this equation analytically. One, advanced by Sips [83, 84], consists in bringing equation (3.138) to the form of the Stieltjes transform [85]. This is how analytical functions of adsorption energy distributions, corresponding to the total adsorption isotherms described by the equations of Freundlich [83, 84, 86, 87] and Dubinin and Radushkevich [87], were determined.

The second analytical method for solving equation (3.138) is simple, yields good results, and is known as the condensation approximation method [88–93]. In this case, the local adsorption isotherm is approximated by the so-called condensation isotherm:

$$\theta_t(p,\varepsilon)=\begin{cases}0 \text{ for } p\leqslant p_c\\ 1 \text{ for } p>p_c\end{cases} \tag{3.139}$$

where p_c is the pressure which initiates condensation of the adsorbed gas (vapour). For the local isotherm (equation (3.139)), equation (3.135) takes the form:

$$\theta_t(p)=\int_{E(p)}^{\infty}\chi(\varepsilon)\,d\varepsilon. \tag{3.140}$$

The adsorption energy distribution function becomes

$$\chi(E)=\frac{d\tilde{\theta}_t(E)}{dE}. \tag{3.141}$$

The relation between energy E and pressure p is determined for a given local adsorption isotherm from the general relation proposed by Cerofolini [92, 93]:

$$\tilde{\theta}_t(E,p)=\tfrac{1}{2} \tag{3.142}$$

whereas the function $\tilde{\theta}_t(E)$ is determined as: $\tilde{\theta}_t(E)=\theta_t[p(E)]$.

A large number of numerical solutions of equation (3.135) are known. However, Adamson's method [94, 95], consisting in the use of successive graphical or analytical approximations, is best known. The results obtained by this method were compared with the $\chi(\varepsilon)$ function obtained calorimetrically by Drain and Morrison [96] for adsorption of argon on rutile (Fig. 3.28). Adamson's method is, however, very laborious.

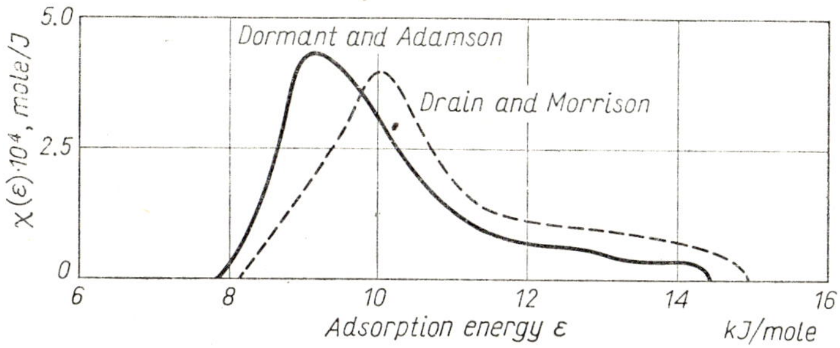

FIG. 3.28 Adsorption energy distribution function of argon on rutile. $\chi(\varepsilon)$ was obtained by the Adamson method [94], and calorimetrically by Drain and Morrison [96].

Other methods of solving equation (3.135) are based on the expansion of the subintegral functions into series of orthonormal functions. The latter methods are much simpler numerically and yield good results [93, 97–100]. The methods of Van Dongen [101], assuming an exponential polynomial for the function $\chi(\varepsilon)$, and of Rudziński *et al.* [102], assuming a linear combination of Freundlich's isotherms for the total adsorption isotherm, also deserve mention.

The effect of adsorption surface heterogeneity on gas (vapour) adsorption may also be studied using a general exponential adsorption isotherm [103, 104]:

$$\theta_t(p) = \exp\left[\sum_{i=1}^{k} B_i\left(RT\ln\frac{p}{p_m}\right)^i\right] \quad \text{for } p \leqslant p_m \tag{3.143}$$

where B_i and p_m are parameters (p_m is the gas pressure for $\theta_t(p)=1$). The Freundlich and Dubinin and Radushkevich adsorption isotherms are special cases. The analytical forms of the adsorption energy distribution function $\chi(\varepsilon)$ corresponding to equation (3.143) may have diverse shapes. The approximate form of the $\chi(\varepsilon)$ function is obtained from equation (3.141) assuming Langmuir's local adsorption model. We then get:

$$\chi(\varepsilon) = \sum_{i=1}^{k} iB_i(E_m - E)^{i-1}\exp\left[\sum_{i=1}^{k} B_i(E_m - E)^i\right] \tag{3.144}$$

where

$$E = RT \ln \frac{K_H}{p} \quad \text{and} \quad E_m = RT \ln \frac{K_H}{p_m}.$$

Experimental studies have confirmed the usefulness of the adsorption isotherm equation (3.143) for the determination of the heterogeneity of the adsorbent surface. In Fig. 3.29 a plot of the adsorption energy distribu-

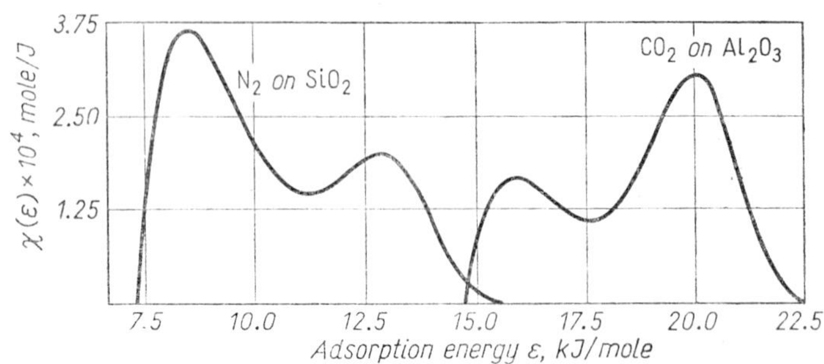

FIG. 3.29 Adsorption energy distribution function for nitrogen on silica gel and CO_2 on alumina (after Jaroniec [105, 106]).

tion function, as calculated using equation (3.144), is shown for the adsorption of nitrogen on silica gel and CO_2 on alumina. In both cases we observe two principal adsorption centre groups for which the distribution of sites is different [105, 106].

3.6.4 Some Problems of Gas (Vapour) Adsorption on Heterogeneous Adsorbent Surfaces

Recent investigations on the effect of adsorbent surface heterogeneity concentrate on two problems:

(i) multilayer adsorption on heterogeneous surfaces, and

(ii) effect of choice of local adsorption model on the form of the adsorption energy distribution function.

The BET adsorption isotherm can be represented as a product of a function $f(p)$ and Langmuir's adsorption isotherm $\theta_L(p, \varepsilon)$ [92]:

$$\theta_w(p) = f(p)\,\theta_L(p, \varepsilon) = \frac{1}{1 - p/p_0} \; \frac{\dfrac{p}{K_H} \exp\left(\dfrac{\varepsilon}{RT}\right)}{1 + \dfrac{p}{K_H} \exp\left(\dfrac{\varepsilon}{RT}\right)} \qquad (3.145)$$

when $C_{BET} \gg 1$. If we use this equation as the local adsorption isotherm in the integral equation, we get:

$$\theta_w(p) = f(p) \int_{\Omega} \theta_L(p, \varepsilon)\chi(\varepsilon)\,d\varepsilon = f(p)\theta_t(p). \tag{3.146}$$

To obtain the equations for the multilayer adsorption isotherm it is necessary to multiply those for monolayer adsorption by the multilayer factor $f(p)$. This is how Cerofolini generalized the D–R adsorption isotherm for multilayer adsorption. These studies were developed by Rudziński *et al.* [107–110], who used the BET equation, assuming the formation of a finite number of layers in the surface phase (see equation (3.66)), as the local adsorption isotherm. Numerical calculations for argon adsorption on rutile [107] revealed that six layers of adsorbate molecules are formed in the surface phase (Fig. 3.30). This result is in agreement with the studies of Brunauer *et al.* [111] who showed that the number of layers formed in the course of multilayer gas adsorption is usually 5 or 6.

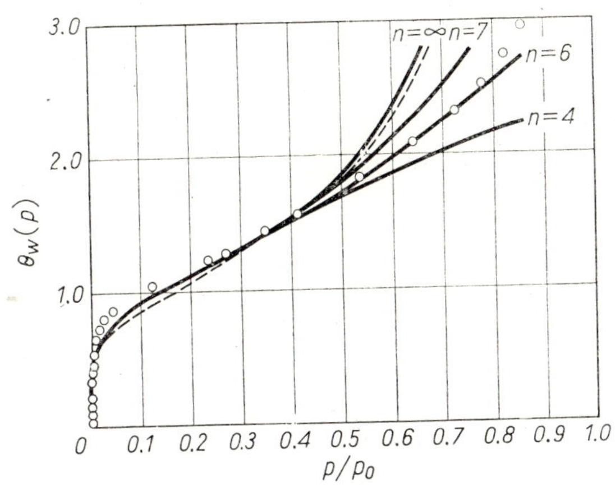

Fig. 3.30 Adsorption isotherms for argon on rutile calculated from equation (3.145) assuming different numbers of layers (n) (after Rudziński *et al.* [107]). Open circles denote experimental points (after Drain and Morrison [96]).

Comprehensive theoretical considerations of the effect on the shape of the adsorption energy distribution function of selecting a given local adsorption isotherm were conducted by Misra [112] for the Jovanovič adsorption isotherm (see equations (3.75) and (3.76)). The application of this equation in the integral equation (3.135) yields the Laplace transform [79, 113]. House and Jaycock [114] applied the above Adamson method to the determination of the $\chi(\varepsilon)$ function. These authors also used the

Langmuir and Hill–de Boer equations as local adsorption isotherms. Interesting results are also obtained using the condensation approximation. The choice of local adsorption isotherm appears to affect only the position of the $\chi(\varepsilon)$ function with regard to the energy axis (Rudziński and Jaroniec [113]). Waksmundzki *et al.* [115, 116], and Cerofolini [117, 118] propose analytical equations which allow an estimate of the shift of the $\chi(\varepsilon)$ function with respect to the energy axis, for any local adsorption isotherm, as compared with the $\chi(\varepsilon)$ function calculated assuming Langmuir's equation as local adsorption isotherm. When determining the heterogeneity of the adsorbent surface, the proper local adsorption model should be assumed (on the homogeneous sector of the surface area), i.e. localized or non-localized. This choice of model can be based on thermodynamic studies [119].

3.7 Some Methods of Measuring Gas and Vapour Adsorption

Isotherms for the adsorption of gases and vapours on solids may be determind by static or dynamic methods. The former consist of an adsorbent placed in the atmosphere of a gas or vapour, and, after equilibrium is established, the pressure and quantity of adsorbate adsorbed or the differences between the adsorbate introduced and that remaining in the gaseous phase (evaporation of the liquid from a vacuum microburette, gas-volume method) are measured. Static methods are usually applied when determining adsorption of individual gases or vapours in a vacuum apparatus in which the adsorbent has been subjected to preliminary heating under high vacuum in order to release the substances earlier adsorbed on its surface.

Figure 3.31 shows the apparatus for measuring vapour adsorption by means of McBain's balance [120]. The pan *3* with a weighed-out portion of adsorbent hangs on a quartz spring *2* in a sealed glass tube *1*. The lower part of the tube is enclosed in a thermostatted box *6*. After introducing the gas (vapour) into the apparatus, equilibrium is established between the gas (vapour) adsorbed and its pressure in the gaseous phase (measured by manometer *5*). The increase in weight of the adsorbent is read from the elongation of the calibrated quartz spring. In the case of vapour adsorption it is very convenient to increase the pressure by connecting an ampoule *4* with the liquid substance, to the apparatus. The ampoule is enclosed in a thermostat box *7* whose temperature controls the pressure of the vapour in the apparatus.

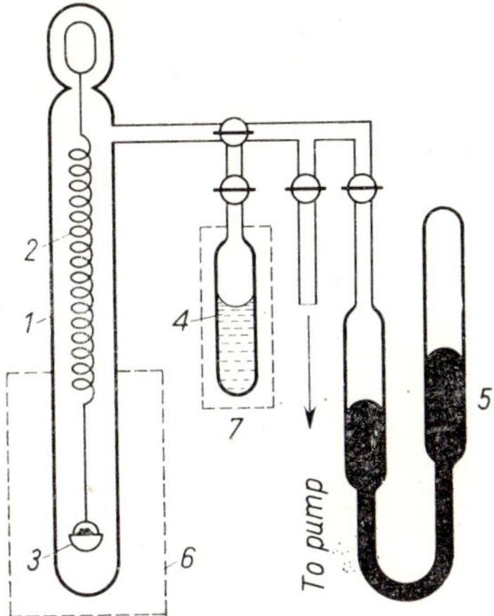

FIG. 3.31 Diagram of apparatus for measuring adsorption of gas or vapour involving McBain's balance (after McBain and Baker [120]).

Figure 3.32 illustrates the apparatus with a vacuum microburette *2* from which the vapour of the liquid substance is fed to the ampoule *1* with the adsorbent. The quantity of adsorbed substance is measured as its loss in the microburette – the quantity of substance remaining in the gas phase being known. The volume of the so-called gas pocket of the apparatus is determined using helium whose adsorption may be neglected.

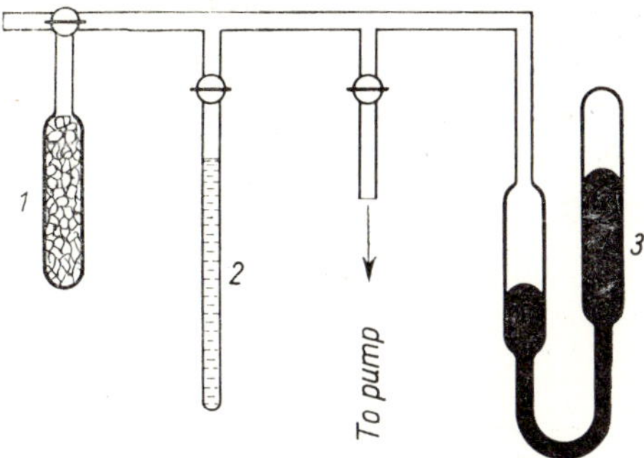

FIG. 3.32 Diagram of apparatus for measuring adsorption of gas or vapour by means of microburette after Dreving and Eltekov [121]).

The equivalent pressure of the vapour adsorbed is measured by means of manometer *3*. In the gas-volume method the principle of adsorption measurement is the same, but the liquid microburette is replaced by a gas microburette.

Dynamic methods of measuring gas adsorption have increased in importance owing to the development of the theory of gas adsorption chromatography. Gas adsorption chromatography has become the fundamental method of studying adsorption of gases and vapours, due to its speed and accuracy.

The chromatographic method consists in passing through a bed (column) of adsorbent a flux composed of a mixture of inert gas (e.g. He or N_2), referred to as carrier gas, and an adequate amount (at suitable partial pressure) of the gas (vapour) to be adsorbed. The gradual increase of gas concentration at the adsorbent column outlet is subsequently analysed (the so-called outlet curves). Another method consists in analysing the variations in concentration of the adsorbed gas washed out from the adsorbent column by the carrier gas.

Since many monographs have recently appeared in the field of gas chromatography [121], we present here only the experimental basis of the method. A diagram of the gas chromatograph is presented in Fig. 3.33.

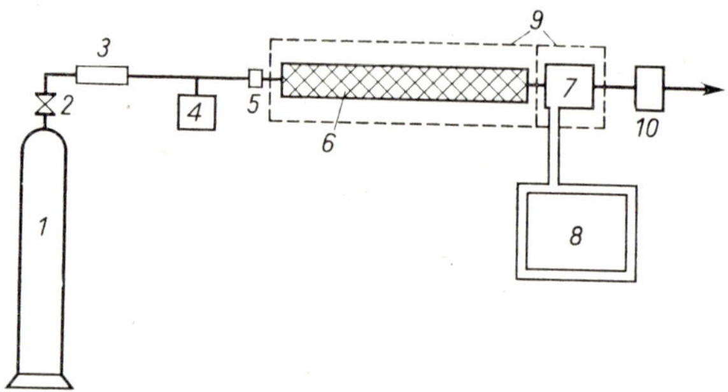

FIG. 3.33 Diagram of gas chromatograph.

The carrier gas passes from cylinder *1* through valve *2*, filters *3* (decontaminating and drying) and flow rate controlling device *5* to the adsorbent-containing column *6*. The pressure of the gas is monitored by manometer *4*. The gas leaving the chromatographic column enters the detector *7*, which monitors the appearance of the substance under investigation; the signal from the detector is transferred after amplification to the autographic recorder *8*. The chromatographic column and detector are enclosed in a thermostatted chamber *9*. The flow rate of the gas at the exit end is measured by means of the flow meter *10*.

Analysis of the curves drawn by the autographic recorder allows us to calculate (under suitably chosen conditions) the adsorption of a given vapour or gas under a given pressure, i.e. the adsorption isotherm. It is also possible to directly calculate the differential heats of adsorption of a gas, for example, from the chromatographic data.

3.8 Adsorbents – their Production and Structure

Many adsorbents of different chemical character and different geometrical surface structure are currently available. Here we shall concentrate on the basic problems relating to the construction and structure of adsorbents, and we shall describe the most important and most commonly used adsorbent types.

3.8.1 Non-porous Adsorbents

Non-porous adsorbents are obtained by precipitation of crystalline deposits such as $BaSO_4$ or by grinding vitreous or crystalline solids. The specific surface area of these adsorbents is small and seldom exceeds 10 m^2/g; more frequently it lies in the range of 0.1 to 1 m^2/g.

Finely divided non-porous materials, e.g. rubber fillers, may be obtained by incomplete combustion of organic substances (carbon black) or organosilicon compounds (so-called white carbon black), as well as by hydrolysis of orthosilicic haloanhydrides ($SiCl_4$, SiF_4) in strongly superheated steam (silica aerogels are thus obtained). The specific surface area of such materials, composed of non-porous particles, attains hundreds of square metres per gram. These adsorbents find wide application as fillers for polymers, lubricants, lakes, etc.

Graphitized carbon blacks are obtained from normal carbon black by heating at 3000°C in vacuum, in an inert gas or reducing atmosphere [122]. At 3000°C the carbon black particles assume the form of polyhedra whose planes are built of graphite. In the case of thermal black the surface consists of the principal planes of graphite. Therefore the adsorption properties of carbon black graphitized at 3000°C closely resemble those of the homogeneous principal surface of the graphite crystals. The specific surface area of graphitized carbon blacks reaches several dozen square metres per gram.

3.8.2 Porous Adsorbents

To ensure the effective operation of adsorbents for the retention of gases and vapours, as active contacts or catalyst supports, for dehydration as well as adsorptive separation of mixture components, solids are used with specific surface areas ranging from several hundred to one thousand square metres per gram. Such adsorbents are used in granular form (tablets, granules, beads) to give them sufficient strength and to reduce the resistance offered to the gas or liquid stream. The mesh size of the grains usually ranges from 0.1 to 2 mm. Such requirements are fulfilled by adsorbents with a sufficiently narrow porous structure.

Porous adsorbents with large specific surface areas can be obtained by various methods of which two are most important.

The first method consists of the construction of a rigid adsorbent skeleton of small, colloidal particles. The latter conglomerate or fuse at the sites of contact yielding a skeleton with a huge inner surface area. Among adsorbents with corpuscular structure are numerous dried gels (xerogels) such as silica gels, aqueous Al_2O_3, activated magnesia, etc. The component particles of such adsorbents may be amorphous, and usually have the shape of globules, e.g. silica and alumino-silica gels. The particles composing the skeleton may also be fine crystals.

The specific surface area of corpuscular adsorbent depends upon the size of particles comprising the skeleton. For instance, the specific surface area of adsorbents composed of adjacent globules of similar sizes is equal to the product $\pi d^2 n$, where d is the diameter and n the number of globules in 1 g of adsorbent. The volume v of 1 g of adsorbent is given by the equation: $v = 1/\rho \approx \frac{1}{6} d^2 n$, where ρ is the particle density. Hence

$$s \approx 6/\rho d \quad \text{and} \quad d \approx 6/\rho s .$$

In this way, from a knowledge of the specific surface area of the adsorbent, we can estimate the mean size of the particles which comprise the skeleton. If $\rho \approx 2.2$ g/cm^3 (silica gels), then $d \approx 2730/s$ nm. In turn, if $s \approx 100$ m^2/g (low activity silica gels), then $d \approx 30$ nm; if $s \approx 800$ m^2/g (high activity silica gels), then $d \approx 4$ nm.

The capillaries (pores) in an adsorbent of corpuscular structure are the free spaces between the particles. The sizes of these capillaries are dependent on the dimensions of the particles and on their packing. When the particle dimensions only differ slightly and their packing is close to ideal, more or less homogeneous porous structures are obtained. The average number of adjacent globules in uniformly porous active silica gels amounts to about six; this corresponds approximately to simple spatial packing (Fig. 3.34).

The second method of obtaining porous adsorbents consists of the exposure of wide-porous or non-porous materials (coke, glass) to the action of active gases or liquids. For instance, when acting on non-active

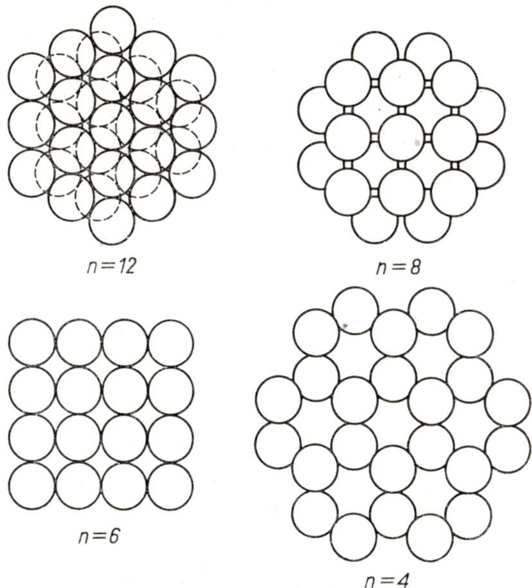

Fig. 3.34 Simple models of uniformly porous silica gel structures with varying numbers of adjacent globules: 4, 6, 8 and 12.

carbon with oxygen-bearing gases (H_2O or CO_2) at 1123–1223 K, part of the carbon undergoes combustion as a result of which we obtain active carbon with a highly developed surface (size of pores from several to several dozen nanometres). Action with acids on sodium–boron glass removes the sodium–boron components, and yields porous glass with pore dimensions similar to those of active carbon; the size of pores is dependent on the heat-conditioning of the glass and final washing with aqueous NaOH and KOH solutions.

In both cases, we obtain porous solids, crystalline or amorphous, with a more or less developed surface area. Such a porous structure, with a given arrangement and orientation of the adjacent particles and open spaces between them, is often called the adsorbent texture.

Adsorbents may have different porosity, where the pores may differ both in size and in shape. However, in many cases the pore width is the most important feature. Assuming pores with cylindrical shape, we speak of their diameter, whereas if the pores are slit-shaped, we can speak of their width. Dubinin [123] has given the following classification of pores

(i) Micropores: diameter <2 nm.

(ii) Mesopores: diameter 2–200 nm.

(iii) Macropores: diameter >200 nm.

Dubinin's classification has been fairly widely accepted in the literature concerned with adsorption and adsorbents. According to de Boer [124] macropores are pores of width greater than 50 nm.

Silica gel is one of the most commonly used adsorbents. The vast literature of this subject [125–130] deals mainly with its production and properties. Silica gel has the general formula $SiO_2 \cdot xH_2O$ and occurs in a wide range of amorphous forms consisting of SiO_4 tetrahedra arranged in a disordered spatial lattice [131, 132] (Fig. 3.35). The properties of this adsorbent depend on the mutual orientation of the edges of surface tetrahedra with which hydroxyl groups are combined (see Chapter 6).

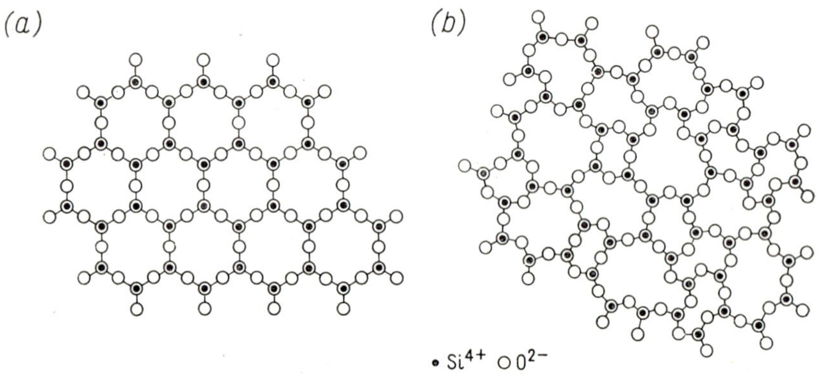

FIG. 3.35 The arrangement of SiO_4 tetrahedra in crystalline (a) and amorphous–vitreous (b) silica gel.

The common usage of various kinds of silica gel is due to the fact that it is easy to produce and its surface properties (geometric structure of the pores and chemical nature of the surface) are apt to modification.

Silica gel is obtained by polycondensation of orthosilicic acid. The monomer and polymers of this acid are obtained during neutralization of soluble silicates (liquid glass) with inorganic acids or by hydrolysis of such compounds as silicon tetrachloride, $SiCl_4$, or tetraethoxysilane, $Si(CO_2H_5)_4$. Silicic acid polycondensation can be represented as follows:

$$(HO)_3SiOH + HOSi(OH)_3 \rightarrow (HO)_3Si{-}O{-}Si(OH)_3 + H_2O$$

or more generally:

$$(SiO_pH_q)_n - SiOH + HOSi - (SiO_rH_t)_m \rightarrow$$
$$\rightarrow (SiO_pH_q)_n - Si - O - Si - (SiO_rH_t)_m + H_2O\,.$$

The above reaction yields macromolecules of silicic acid which gradually grow (the silicon–oxygen chains grow longer, branch and undergo cyclization). The viscosity of the solution gradually increases, and finally a silicate sol is produced. The sol particles assume spherical shape when their surface area is smallest. The size of the sol particles varies, depending on the process parameters, from 2 to 15 nm. The gradually increasing viscosity of concentrated silicate sols, due to the formation of larger and larger aggregates of sol particles as a result of coupling of surface hydroxyls, leads to the formation of silica hydrogel, which, on drying, yields the xerogel.

A silica gel of definite texture may be obtained in different ways. Systematic studies have shown [134–138] that a relation exists between the structure of silica gel and its method of production. It was found that the rate of polycondensation of silicic acids, which is largely affected by the pH of the reaction medium, is of primary importance. The texture of

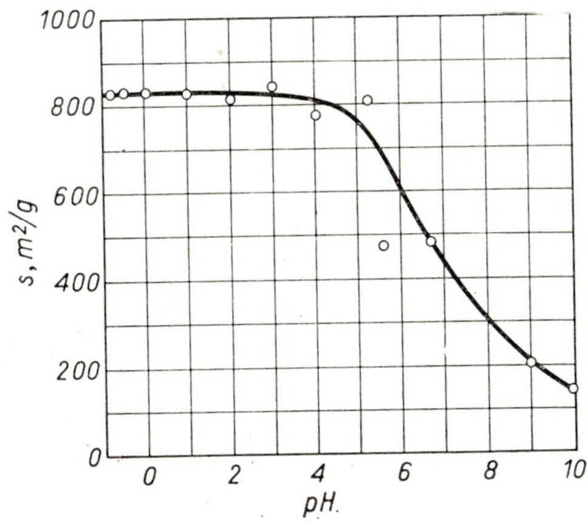

FIG. 3.36 Dependence of the specific surface area of silica gel on the pH of the silicic acids condensation medium (after [133–138]).

adsorbents is characterized primarily by their specific surface area and pore volume. In Figs. 3.36 and 3.37 such variations are presented for silica gel with the pH of the silicic acid condensation medium.

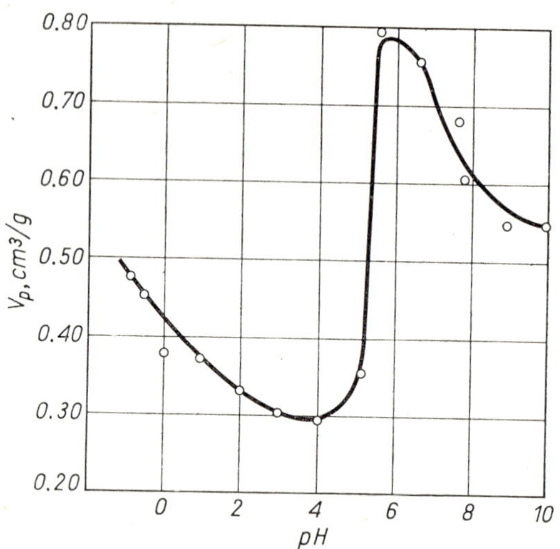

FIG. 3.37 Dependence of the pore volume V_p in silica gel on the pH of the silicic acids condensation medium (after [133–138]).

The geometrical structure of xerogel can also be modified by subjecting hydrogel, prior to drying, to treatment with solutions of salts, dehydrating substances [137] or surface active agents [138]. The packing of the particles in the xerogel skeleton changes with change of surface forces: if they are decreased, xerogel is obtained with wider pores whereas their increase yields a narrow-pore xerogel. Silica gels of specific surface area ranging from several dozen to several hundred square metres per gram can thus be obtained.

Silica gels with very wide pores are obtained by hydrothermal treatment of narrow-porous xerogels. This treatment consists in exposing the gel in an autoclave to the action of steam at high temperature [137, 138]. In this way macroporous silica gels with specific surface area ranging from about 5 to 10 m^2/g are obtained.

Alumina (*active alumina*). This term encompasses a wide range of compounds containing, in addition to Al_2O_3, different quantities of water and often alkali metals or alkali earth oxides.

The first systematic classification of aluminas was proposed by Ginsberg *et al.* [140] and was subsequently modified by Lippens [139]. It is based on the temperature ranges in which alumina is obtained from hydroxides.

(i) Group γ encompasses low-temperature aluminium oxides ($Al_2O_3 \cdot xH_2O$, where $x=0$ to 0.6) obtained at temperatures below 600°C. This group includes ρ-, χ-, η- and γ-aluminium oxides.

(ii) Group δ encompasses high-temperature aluminium oxides (practically anhydrous Al_2O_3) obtained in the temperature range from 900 to 1000°C. This group includes κ-, θ- and δ-aluminium oxides.

Besides this classification, α-alumina, obtained at 1200°C and higher, deserves mention.

Krischner [141] provided a classification of aluminas based on their crystallographic structure, which is related to the extent of packing of the crystallographic lattice elements. However, when considering active kinds of alumina, the classification given by Lippens is most frequently used. Aluminas are chiefly obtained by dehydration of various aluminium hydroxides.

Aluminium hydroxides. Hydrolysis of aluminates or addition of alkali metal carbonates or hydroxides to aluminium salt solutions leads to the formation of a substance known as aluminium hydroxide. More exhaustive studies have shown that the following alumina hydrates occur [126]:

Gibbsite or hydrargillit, Bayerite, Nordstrandite } having the general formula $Al_2O_3 \cdot 3H_2O$

Diaspore, Boehmite } having the general formula $Al_2O_3 \cdot H_2O$

These hydrates are obtained by suitably varying the conditions (temperature, pH and concentration) of aluminate or aluminium salt hydrolysis.

Current opinions on the formation of different forms of alumina in the course of dehydration of hydroxides are not in full agreement [142]. Nevertheless, schemes for that dehydration may be presented under various conditions, e.g.:

(i) Dehydration in air

$$\text{Gibbsite} \xrightarrow{250^\circ C} \chi \xrightarrow{900^\circ C} \kappa \xrightarrow{1200^\circ C} \alpha\text{-}Al_2O_3$$

$$\text{Gibbsite} \;\Big|\; \xrightarrow[180^\circ C]{180^\circ C} \text{boehmite} \xrightarrow{450^\circ C} \gamma\text{-}Al_2O_3$$

$$\left.\begin{matrix}\text{Bayerite}\\ \text{Nordstrandite}\end{matrix}\right\} \xrightarrow{230^\circ C} \eta \xrightarrow{850^\circ C} \theta \xrightarrow{1200^\circ C} \alpha\text{-}Al_2O_3$$

Gel boehmite (pseudo boehmite) $\xrightarrow{300°C} \gamma \xrightarrow{900°C} \delta \xrightarrow{1000°C} \theta + \alpha \xrightarrow{1200°C} \alpha\text{-}Al_2O_3$

(ii) Dehydration in vacuum

$$\left.\begin{array}{l}\text{Gibbsite}\\ \text{Bayerite}\\ \text{Nordstrandite}\end{array}\right\} \xrightarrow{200°C} \rho \text{ or } \eta \xrightarrow{750°C} \theta \xrightarrow{1200°C} \alpha\text{-}Al_2O_3$$

$\xrightarrow{180°C}$ boehmite

The form of alumina most commonly used as an adsorbent is γ-Al_2O_3, with specific surface area of 100–200 m^2/g. According to Lippens [139] and Peri [143], γ-Al_2O_3 has the structure shown in Fig. 3.38. In terms of this model the surface is built of oxygen O^{2-} and Al^{3+} ions, with only 3/4 of the positions specific for aluminium being occupied, the balance constituting surface defects.

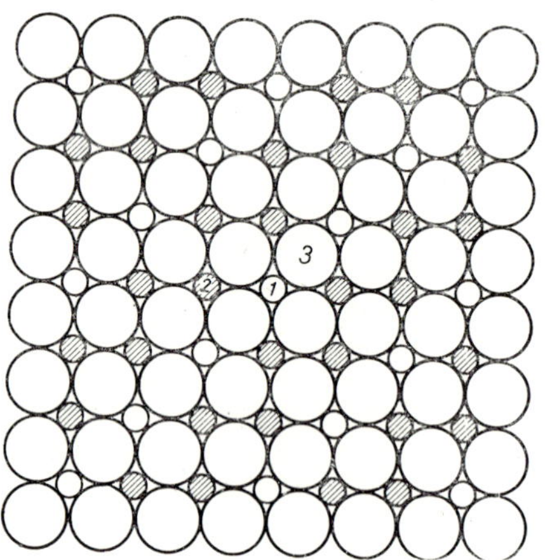

Fig. 3.38 Model of γ-Al_2O_3 surface structure: *1*—surface defects, *2*—Al^{3+} ions, *3*—O^{2-} ions (after Lippens [139]).

Alumina may contain different quantities of surface hydroxyls or adsorbed water. Many authors [144, 145] studying the microporous structure of various aluminas have found that, in addition to regular, cylindrical micropores 2.9 nm in diameter, irregular pores of larger diameter also occur.

Active carbons, the oldest adsorbents in wide use today, are obtained most frequently by removing bituminous substances from crude coal and by partial combustion under the action of H_2O and CO_2 at high temperature. Active carbons can also be obtained by impregnation of various organic materials with such salts as K_2S, $ZnCl_2$, etc. following by roasting in vacuum and washing of the resulting carbon with water.

Active carbons belong to the group of graphitic materials. Their elementary crystals are composed of irregularly arranged six-membered carbon rings, whose crystallites attain dimensions of 2.3×0.9 nm. The skeleton of active carbons is composed of these irregularly interconnected crystallites. Depending upon the mutual arrangement of the crystallites, only part of the inner surface of the active carbon, reaching 2000 m^2/g, is accessible to adsorption. Thus the specific surface area of active carbons, determined primarily by the micropores, lies within the range 400 to 900 m^2/g.

The surface of active carbons is geometrically and chemically heterogeneous. Various groups such as compounds containing hydrogen, oxygen, etc. may be attached to the carbon lattice at the edges of the six-membered rings, depending on the method of production and treatment of the carbon.

Dubinin [146] distinguishes two extreme types of active carbons:

(i) Carbons in which pores 1–3 nm in diameter prevail.

(ii) Carbons which have both micropores and pores with diameters greater than 3 nm.

Molecular sieves. So-called porous crystals which act as molecular sieves are very important, particularly zeolites, which are alkali metal or alkaline earth aluminosilicates consisting of spatially arranged SiO_4 and AlO_4 tetrahedra. The alumino-oxygen tetrahedra differ from the silicon-oxygen ones in electric charge. Substitution of a SiO_4 group an AlO_4 introduces into the crystal lattice an extra negative charge which is neutralized by the Na^+, K^+, Li^+, Ca^{2+}, etc. cations located in the lattice and revealing a certain mobility.

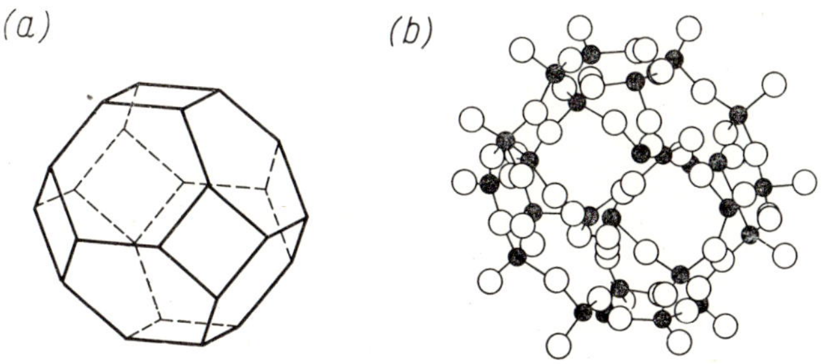

FIG. 3.39 Spatial octahedra: (a) model, (b) spatial arrangement of tetrahedra to form an octahedron (open circles denote oxygen atoms, full circles denote silicon and aluminium atoms).

Twenty-four SiO_4 and AlO_4 tetrahedra form spatial octahedral units (Fig. 3.39) interconnected in the zeolite crystals to form a lattice. Thus the spatial octahedra (cubo-octahedra) whose vertexes are the Si^{4+} or Al^{3+} ions, constitute the structural elements of these lattices.

The general oxide formula of zeolites can be written in the form:

$$Me_{2/n} \cdot Al_2O_3 \cdot mSiO_2 \cdot pH_2O$$

where Me is a metal, n is its valency, m and p are numbers characteristic for the given zeolite.

So far some 40 natural and as many synthetic zeolites are known, the most common and best known of which are zeolites of types A, X and Y.

The zeolite cubo-octahedra have six four-membered rings of octahedral symmetry and eight six-membered rings composing two groups of tetrahedral symmetry on their surface (Fig. 3.39).

In A-type zeolites the single cubo-octahedron is composed of 12 SiO_4 and 12 AlO_4 tetrahedra and a sufficient number of metal cations. In the zeolite A skeleton each cubo-octahedron is connected to neighbouring cubo-octahedra through four-membered rings. Eight aluminosilicate cubo-octahedra interconnected in this way enclose a spherical space into which the molecules of the adsorbate can penetrate in adsorption via six eight-membered windows (Fig. 3.40a).

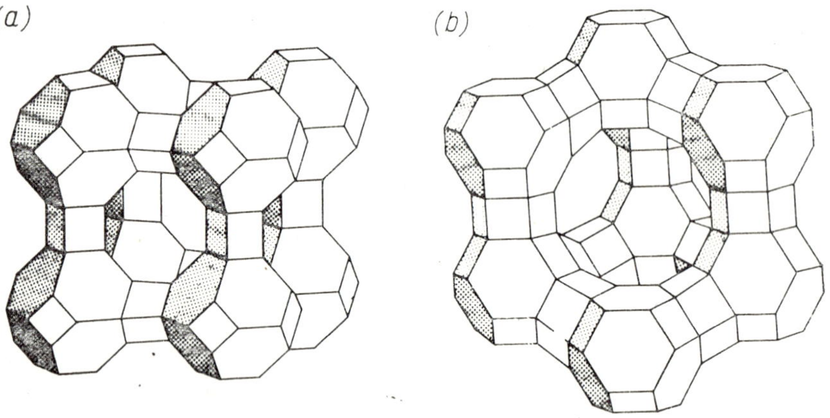

FIG. 3.40 Structural models of zeolites A(a) and X(b).

In the X-type zeolite the cubo-octahedra consist of 14 SiO_4 and 10 AlO_4 tetrahedra, each cubo–octahedron being connected to four others by six-membered rings. Eight cubo-octahedra interconnected in this way enclose a spherical space with four twelve-membered windows (Fig. 3.40b).

The crystallographic lattice structure of zeolite Y is similar to that of zeolite X; it includes, however, more SiO_2 and thus shows higher resistance to the action of temperature, superheated steam, acids and bases. Its increased resistance makes it very useful in industrial applications.

The size of windows leading to the inner spaces of zeolite crystal lattices is also affected by the charge and dimensions of the cations included in these lattices. In the case of the sodium form of zeolite A (4A) the diameter of the windows is about 0.42 nm, in the case of the calcium form (5A) about 0.48 nm, and in the case of the potassium-sodium form (3A) about 0.38 nm. In the sodium form of zeolite X (13X) the diameter of the windows is 0.8–1 nm, whereas in the calcium form (10X) it is about 0.7 nm. The situation is similar for the various forms of zeolite Y.

According to Meier's classification [147] the known zeolites are classified into seven groups. Zeolites A, X and Y are in this classification in the so-called foyasite group.

Zeolites are most frequently synthesized by the hydrothermal method at temperatures up to 100°C under atmospheric pressure, when aluminosilicate gels are crystallized in an alkaline medium. Depending upon the type of zeolite being produced, the composition of the mixture subjected to hydrothermal crystallization (for several hours to several days) is changed accordingly. A white, fine-crystalline powder is obtained, with particle size of 0.5–5 μm, which requires pressing into granules with a bonding agent (bentonite clay added in a proportion of 15–20%) before use. Alongside the initial (intracrystalline) system of pores related directly to the zeolite structure a system of secondary (intercrystalline) pores exists in the granules which appear in the granulation process (bonding of zeolite crystals or crystal aggregates with the bonding agent).

The adsorption capacity of the molecular sieve is of course dependent on the initial system of pores. The adsorption rate is controlled, however, by both systems of pores.

Zeolites operate as molecular sieves via the penetration or non-penetration through the windows of the molecules of various adsorbates, having different sizes and structure, into the inner spaces of the crystalline lattice of the sieves. Of course, only those adsorbate molecules which penetrate into the zeolites are arrested.

Porous glasses. These are rapidly gaining significance as adsorbents, and are obtained from sodium-boron-silicon glass by suitable heat treatment and etching with acids, when the Na_2O and B_2O_3 oxides are dissolved. The Si—O—Si bonds in the silicon–oxygen lattice resist splitting, and the silica skeleton constitutes the basic matrix of porous glass. The structure depends on that of the initial glass and the conditions of its heat treatment. Changes in composition of the initial glass and its heat treatment make it possible to obtain glasses with pores of required diameter. By varying the time of etching with acid solution we can control the depth of the pores in the glass, the mean dimensions of the pores remaining unchanged. To obtain wider pores, the glass obtained by etching with acid may be additionally treated for 1–2 hours with a 0.5 M KOH solution at 293 K. Finely divided silica is then dissolved, the total volume of the pores increases, while the specific surface area decreases. The size (width) of the pores in porous glasses ranges from several tenths to several hundred nanometres.

The possibility of controlling the pore structure in porous glasses permits their use for analytical purposes (chromatography), for the separation of mixtures and purification of substances. Porous glasses with very narrow pores ($\sim$1 nm) can also be used as molecular sieves alongside zeolites.

The chemical behaviour of the surfaces of carefully washed porous glasses is the same as that of silica gels.

3.9 Study of Adsorbent Structure by Adsorption Methods

There are many methods of determining the structure of adsorbents—their specific surface area and porosity—by adsorption methods. Some were discussed in this chapter in the context of the Langmuir, BET and H–J theories of gas and vapour adsorption. Such methods lead to the determination of the specific surface area, with the proviso, however, that the size of the molecules of the adsorbed gas is in each case smaller than the capillary size.

The porosity of adsorbents, i.e. the size and volume of their capillaries, can be studied by capillary condensation. Since the meniscus which appears in desorption is spherical, while in adsorption it can be both spherical and cylindrical, the desorption branch of the isotherm may permit measurement of the effective capillary size, i.e. size corresponding to the cylindrical capillaries. Every point of the desorption branch of the hysteresis loop defines the adsorption a and corresponding pressure p/p_0. By multiplying a by V_m (molar volume of the liquid) we obtain the volume V_p of capillaries filled with liquid. From the pressure p/p_0 corresponding to that volume and Kelvin's equation we find the effective radius r_{sph} of the spherical meniscus in the capillary (the wetting angle $\theta=0$):

$$r_{\text{sph}}=\frac{-2\gamma V_{\text{m}}}{RT\ln\dfrac{p}{p_0}}\,. \tag{3.147}$$

The effective capillary radius r is found by adding the thickness l^{s} of the adsorption layer to r_{sph}, viz.

$$r=r_{\text{sph}}+l^{\text{s}}.$$

The thickness l^{s} can be found from the vapour adsorption isotherm of the same substance on a non-porous or very wide-porous adsorbent of the same chemical nature. From a number of V_{p} and r values, we can plot the V_{p} vs. r structural curve for the given adsorbent (Fig. 3.41). By graphical

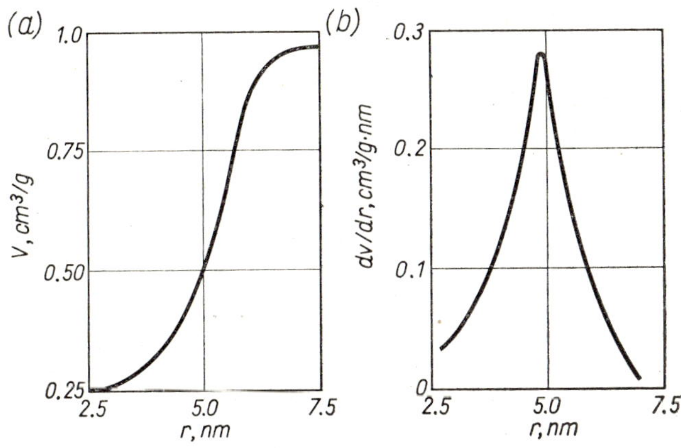

FIG. 3.41 Structural curve (a) and capillary volume distribution curve according to their effective radii (b).

differentiation of this curve, we obtain the capillary volume distribution curve according to the effective radii, dv/dr vs. r, allowing us to determine the porosity of the adsorbent under investigation (Fig. 3.41b). This distribution curve shows that the adsorbent is uniformly porous, and an effective radius of 5 nm is most likely.

For the rapid estimation of adsorbent structure, we can use a series of uniformly porous adsorbents, with pore size (radius) varying from very large (dozens of nanometres) to very small, close in size to that of many hydrocarbon molecules.

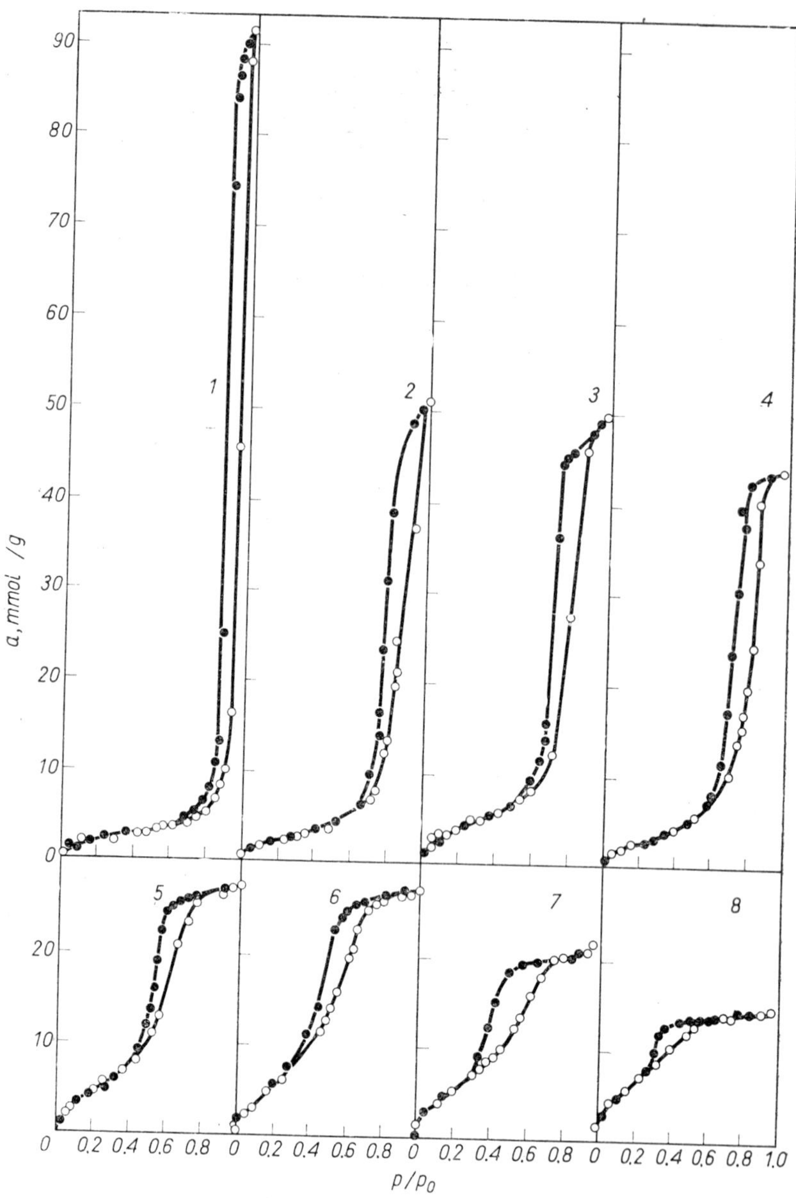

FIG. 3.42 Water vapour adsorption–desorption isotherms for the reference series of silica gels (after Neimark and Sheinfain [138] and Dzhygit *et al.* [148]).

Kiselev, Neimark *et al.* [138, 148] selected eight reference adsorbents from a large number of different silica gels. The corresponding water vapour adsorption–desorption isotherms are shown in Fig. 3.42. All reveal capillary condensation hysteresis loops with steeply descending adsorption branches, thus making it possible to determine the capillary size from the Kelvin equation. The effective diameters d of these capillaries lie in the range of 21.0–2.2 nm. Figure 3.43 presents the capillary volume distribution curves

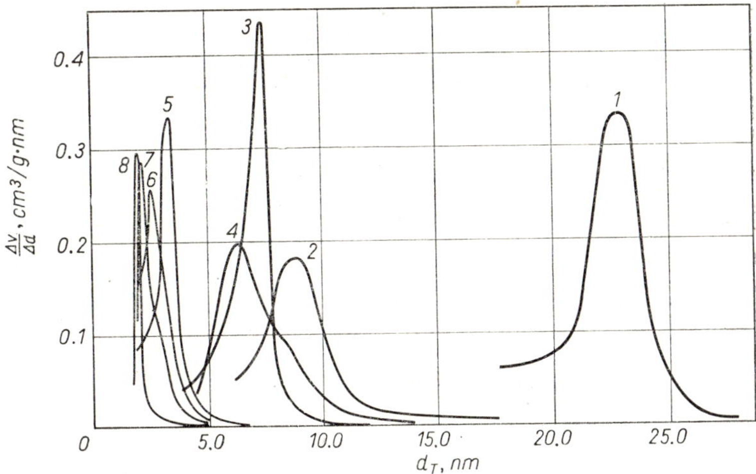

FIG. 3.43 Capillary volume distribution curves for the reference series of silica gels according to their diameters d_T (after Neimark and Sheinfain [138], and Dzhygit *et al.* [148]).

for the reference series of adsorbents according to capillary diameter d_T as calculated from the Kelvin equation (effective capillary diameter $d = d_T + 2l^s$). The specific surface areas of the reference adsorbents were measured by the BET method using nitrogen vapour adsorption isotherms at the boiling temperature. The main structural characteristics of reference series of silica gels are given in Table 3.2.

From the data in Figs. 3.42 and 3.43, as well as in Table 3.2, the structure of the silica gel or adsorbent of similar chemical nature under investigation can be estimated with reasonable accuracy. It suffices to measure the water vapour adsorption–desorption isotherm for the given adsorbent and compare it with the isotherms presented in Fig. 3.42.

To investigate the structure of microporous adsorbents, we can use the Dubinin–Radushkevich adsorption isotherm representing the experimental

Table 3.2 Structural characteristics of the silica gel reference series (after Dzhigit *et al.* [148])

No.	Total volume, V_p, of capillaries cm³/g	Specific surface area m²/g	Capillary diameter, *d* nm	Particle diameter, *D* nm
1	1.70	315	21.0	8.7
2	0.92	350	10.2	8.0
3	0.90	430	8.2	6.3
4	0.80	390	7.5	7.0
5	0.49	575	3.8	4.7
6	0.49	625	3.1	4.5
7	0.39	490	2.9	5.6
8	0.25	540	2.2	5.1

data in the co-ordinate system given in Fig. 3.8 (p. 41). This procedure allows us to determine the micropore volume in the low pressure range. By selecting adsorbates with varying particle sizes, we can estimate the distribution of pores according to their size.

3.10 Chemisorption from the Gaseous Phase

By chemisorption we understand the process of adsorption connected with the formation of chemical bonds between the adsorbate and the solid surface. The chemical bonds formed in chemisorption do not differ in principle from normal chemical bonds. However, many surface compounds have no equivalents in the chemistry of three-dimensional compounds, since the adsorbent atoms participating in chemisorption are at the same time components of the crystalline lattice.

Chemisorption is strongly localized, limited to the monomolecular surface layer, and is related to a heat of adsorption ranging from 80 to 650 kJ/mole.

Chemisorption is usually activated adsorption, proceeding only after preliminary activation of the system. The identification of chemisorption with activated adsorption is therefore not quite correct, since an activation energy is not specific to chemisorption, as chemisorption can proceed without such an activation energy.

The above features of chemisorption are in good agreement with the assumptions of the Langmuir adsorption theory, and the Langmuir isotherm is often used to describe chemisorption.

The assumption of localized positions of adsorbate molecules on the surface and of the monomolecular surface layer simplifies the picture of chemisorption as compared with physical adsorption. However, in this case the heterogeneity of the solid surface becomes very significant.

Langmuir's adsorption model assumes the equivalence of active centres on the adsorbent surface and the independence of adsorption energy on coverage θ. Eventually, however, the active centres are not equivalent and particular molecules are chemisorbed with different energy. The extension of Langmuir's adsorption model thus assumes a dependence of adsorption energy on surface coverage. The Langmuir adsorption isotherm then becomes:

$$\theta = \frac{kp}{1+kp}. \tag{3.148}$$

In Section 3.5 it has been shown that:

$$k = g_0 \exp \frac{q_d}{RT} \tag{3.149}$$

where g_0 is the entropy coefficient and q_d is the differential molar heat of adsorption. Hence:

$$\frac{\theta}{1-\theta} = p g_0 \exp \frac{q_d}{RT}. \tag{3.150}$$

In the case of chemisorption one can of course assume various types of dependence of q_d on the coverage of the surface. Temkin [149] proposed the following dependence:

$$q_d = q_d^0(1-a\theta) \tag{3.151}$$

where a is a constant. After substitution into equation (3.150) we get:

$$\frac{\theta}{1-\theta} = g_0 p \exp \frac{q_d^0(1-a\theta)}{RT}. \tag{3.152}$$

After taking logarithms we can write:

$$\ln p = -\ln C_0 + \frac{a\theta q_d^0}{RT} + \ln \frac{\theta}{1-\theta} \tag{3.153}$$

where $C_0 = g_0 \exp (q_d^0/RT)$. For values of θ close to 0.5 we can write:

$$\theta = \frac{RT}{aq_d^0} \ln C_0 p. \tag{3.154}$$

Equation (3.154), giving the dependence of coverage θ on the logarithm of gas pressure p, is referred to as Temkin's adsorption isotherm. This equation has been found to adequately describe many chemisorption processes [150].

In many cases, Freundlich's adsorption isotherm may be used to describe and interpret chemisorption. Both the Temkin and Freundlich adsorption isotherms adequately describe experimental data, especially in the case of medium surface coverage.

3.11 Kinetics of Adsorption from the Gaseous Phase

The rate of gas adsorption on the homogeneous surface of non-porous or wide-porous adsorbents is the resultant of their adsorption and desorption rates. The quantity a_t of the substance adsorbed by 1 g of adsorbent in time t from the beginning of the process is then given by the equation:

$$a_t = a(1 - \mathrm{e}^{-k_a t}) \tag{3.155}$$

and the rate of adsorption is:

$$\frac{\mathrm{d}a_t}{\mathrm{d}t} = k_a a \mathrm{e}^{-k_a t} \tag{3.156}$$

where a is the equilibrium amount of adsorbed substance as $t \to \infty$, k_a is the adsorption rate constant, dependent on the nature and surface of the adsorbent, as well as on the temperature and pressure of the gas. Figure

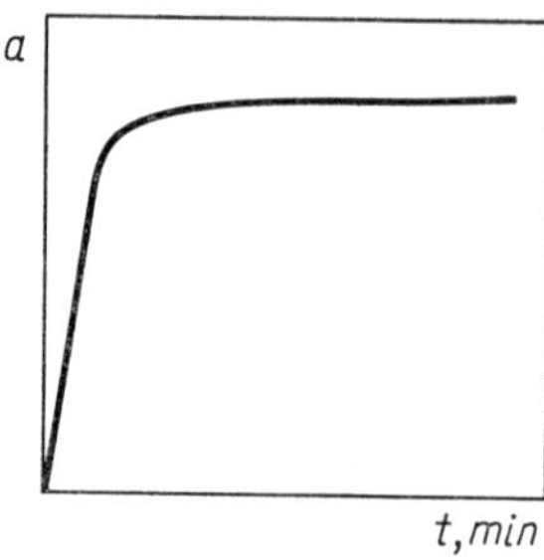

FIG. 3.44 Adsorption rate curve, i.e. kinetic adsorption curve.

3.44 shows the curve described by equation (3.155) (a particular case of the so-called kinetic curve). Although $a_t = a$ for $t \to \infty$, a state not differing appreciably from equilibrium is established after 10^{-5} to 10^{-6} s (under normal pressure).

Attempts have been made to use the first order kinetic equation to describe the rate of gas adsorption on porous adsorbents:

$$\frac{da_t}{dt}=k_a(a-a_t). \tag{3.157}$$

The constant k_a depends on the diffusivity D of the gas into the adsorbent pores. This diffusion is a much slower process than proper adsorption. Transfer of the substance proceeds in the gas phase into the pore (molecular and Knudsen's diffusion) and on to the adsorbent surface (surface diffusion).

The magnitude of effective diffusivity D depends on the character of the porosity of the adsorbent and on that of its surface. It varies in a wide range, from 10^{-3} cm^3/s for wide-porous adsorbents to 10^{-8} cm^3/s for zeolites.

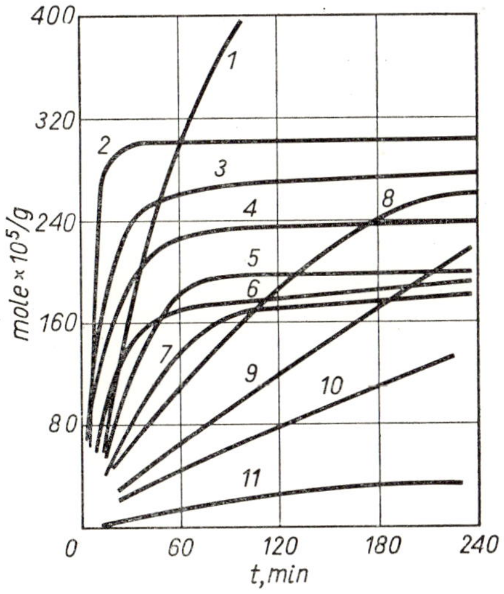

FIG. 3.45 Adsorption rate curves of saturated vapours on carbon (after Trykorn and Wyatt [151]): *1*–CH_3OH, *2*–CS_2, *3*–CH_3COCH_3, *4*–CH_3COOCH_3, *5*–C_2H_5I, *6*–$CHCl_3$, *7*–C_6H_6, *8*–C_2H_5OH, *9*–H_2O, *10*–C_3H_7OH, *11*–C_4H_9OH.

It appears that it will be difficult to find an equation for the adsorption rate which takes account of all parameters, such as the heat effect causing the rise in temperature of the surface layer in the first stage of the adsorption process or the change in the diffusion rate due to change of concentration gradient.

For the above reasons, plots representing the variation of amount of substance adsorbed with time of adsorption afford the most suitable treatment to date of the experimental data for the kinetics of adsorption (Fig. 3.45) [151].

References

[1] Brunauer, S., *The Adsorption of Gases and Vapours*, Princeton University Press, Princeton–New York 1945.

[2] Young, D. M. and Crowell, A. D., *Physical Adsorption of Gases*, Butterworth and Co. (Publishers) Ltd., London 1962.

[3] Kalberer, W. and Mark, H., *Z. physik. Chem.*, **A 139**, 151 (1928).

[4] Magnus A. and Kalberer, W., *Z. anorg. Chem.*, **164**, 357 (1927).

[5] Chaplin, R., *Phil. Mag.*, (7) **2**, 1198 (1926).

[6] Weisman [illegible] and Neumann, W., *Z. Pflernähr. Düng.*, **40,** 49 (1935).

[7] Lambert, H and Peel, D. H. P., *Proc. Roy. Soc.*, **A 144**, 205 (1938).

[8] Emmett, P. H., Brunauer, S. and Love, K., *Soil. Sci.*, **45**, 57 (1938).

[9] Boedecker, C., *J. Landw.*, **7,** 48 (1895).

[10] Freundlich, H., *Colloid and Capillary Chemistry*, Methuen, London 1926.

[11] Halsey, G. D. *Advances in Catalysis*, Vol 4, W. G. Frankenburg, V. I. Komarewski and E. K. Rideal (Eds.), Academic Press, New York 1952, p. 259.

[12] Appel, J., *Surface Sci.*, **39**, 237 (1973).

[13] Langmuir, I., *J. Am. Chem. Soc.*, **38**, 2221 (1916); **40**, 1361 (1918).

[14] Williams, A. M., *Proc. Roy. Soc. (Edinburgh)*, **38**, 23 (1918); **39**, 48 (1919); *Proc. Roy. Soc.*, **A 96**, 287, 298 (1919).

[15] Fowler, R. H. and Guggenheim, E. A., *Statistical Thermodynamics*, Cambridge University Press, Cambridge 1939.

[16] Volmer, M., *Z. physik. Chem.*, **115,** 253 (1925).

[17] Schay, G., *Acta Chim. Acad. Sci. Hung.*, **3**, 511 (1953); *J. chim. phys.*, **53**, 691 (1956).

[18] Eucken, A., *Verh. Deutsch. Phys. Ges.*, **16**, 345 (1914).

[19] Polanyi, M., *Verh. Deutsch. Phys. Ges.*, **16**, 1012 (1914); **18,** 55 (1916); *Z. Elektrochem.*, **26,** 371 (1920).

[20] Titoff, A., *Z. physik. Chem.*, **74**, 641 (1910).

[21] Berenyi, L., *Z. physik. Chem.*, **94**, 628 (1920).

[22] Dubinin, M. M., Zaverina, E. D. and Radushkevich, L. W., *Zh. Fiz. Khim.*, **21**, 1351 (1947).

[23] Dubinin, M. M. and Zaverina, E. D., *Zh. Fiz. Khim.*, **29**, 1129 (1949).

[24] Dubinin, M. M., *Zh. Fiz. Khim.*, **39**, 1305 (1965).

[25] Radushkevich, L. W., *Zh. Fiz. Khim.*, **23**, 1410 (1949).

[26] Dubinin, M. M and Zaverina, E. D., *Dokl. Akad. Nauk SSSR*, **72**, 319 (1950); Dubinin M. M. *Proc. Conf. Ind. Carb. Graph.*, p. 219, S. C. I. (1958).

[27] Brunauer, S., Emmett, P. H. and Teller, E. J., *J. Am. Chem. Soc.*, **60**, 309 (1938).

[28] Hill, T. L., *J. Chem. Phys.*, **14**, 263 (1946).
[29] Pickett, G., *J. Am. Chem. Soc.*, **67**, 535 (1946).
[30] Smith, R. N. and Pierce, C., *J. Phys. Chem.*, **52**, 1115 (1948).
[31] Hüttig, G. F., *Monatsh.*, **78**, 177 (1948).
[32] Ross, S., *J. Phys. Chem.*, **53**, 383 (1949).
[33] Kiselev, A. W., *Koll. Zh.*, **20**, 338 (1958).
[34] Hill, T. L., *J. Chem. Phys.*, **14**, 441 (1946).
[35] De Boer, J. H., *Dynamic Character of Adsorption*, Clarendon Press, Oxford 1953.
[36] Hill, T. L., *J. Chem. Phys.*, **15**, 767 (1947); **16**, 181 (1948); **17**, 772 (1949).
[37] Harkins, W. D. and Jura, G., *J. Am. Chem. Soc.*, **66**, 1366 (1944).
[38] Jura, G and Harkins, W. D., *J. Am. Chem. Soc.*, **68**, 1941 (1946).
[39] Davis, R. T. and de Witt, T. W., *J. Am. Chem. Soc.*, **70**, 1135 (1948).
[40] Livingston, H. K., *J. Chem. Phys.*, **12**, 466 (1944).
[41] Emmett, P. H., *J. Chem. Soc.*, **68**, 1784 (1946).
[42] Loeser, E. H. and Harkins, W. D., *J. Am. Chem. Soc.*, **72**, 3427 (1950).
[43] Jovanovič D. S., *Kolloid Z. u. Z. Polym.*, **235**, 1203 (1970).
[44] Jovanovič D. S., *Kolloid Z. u. Z. Polym.*, **235**, 1214 (1970).
[45] Thomson, W., *Phil. Mag.*, **42**, 448 (1871).
[46] Cohan, L. H., *J. Am. Chem. Soc.*, **60**, 433 (1938).
[47] Isirikyan, A. A. and Kiselev, A. V., *Dokl. Akad. Nauk SSSR*, **110**, 1009 (1956).
[48] De Boer, J. H., *The Structure and Properties of Porous Materials*, Proc. X Symp. Colston Research Soc. Univ. Bristol, Butterworths Sci. Publ., London 1958, p. 68.
[49] Kiselev, A. V. and Pavlova, L. F., *Kinetika i kataliz*, **2**, 599 (1961).
[50] Zhdanov, S. P., Kiselev, A. V. and Pavlova, L. F., *Kinetika i kataliz*, **3**, 445 (1962).
[51] Kiselev, A. V., *Dokl. Akad. Nauk SSSR*, **49**, 516 (1945); *Zh. Fiz. Khim.*, **20**, 239 (1946); *Usp. Khim.*, **15**, 456 (1946).
[52] Hill, T. L., *J. Chem. Phys.*, **17**, 507, 520 (1949); **18**, 246 (1950).
[53] Hill, T. L., *Trans. Faraday Soc.*, **47**, 376 (1951).
[54] Everett, D. H., *Trans. Faraday Soc.*, **46**, 453, 942, 957 (1950).
[55] Everett, D. H. and Young, D. M., *Trans. Faraday Soc.*, **48**, 1164 (1952).
[56] Schay, G., *J. Colloid Int. Sci.*, **35**, 254 (1971).
[57] Hill, T. L., *Advances in Catalysis*, **4**, 212 (1952).
[58] Everett, D. H., *Proc. Chem. Soc.* (*London*), 38 (1957).
[59] Innes, W. B. and Rowley, H. H., *J. Phys. Chem.*, **49**, 411 (1945).
[60] Innes, W. B. and Rowley, H. H., *J. Phys. Chem.*, **51**, 1154 (1947); **55**, 1324 (1951).
[61] Guggenheim, E. A., *Thermodynamics*, North-Holland Publ. Co., Amsterdam 1957.
[62] Manual of Definitions, Terminology and Symbols in Colloid and Surface Chemistry, *Pure and Appl. Chem.*, **31**, 578 (1972).
[63] Kington, G. L. and Aston, J. G., *J. Am. Chem. Soc.*, **75**, 1929 (1951).
[64] Bezus, A. B., Kiselev, A. V. and Sedlaček, Z., *Zh. Fiz. Khim.*, **43**, 1223 (1969).
[65] Isirikyan, I. I., *Metody issledovanya struktury vysokodispersnykh i poristykh*

tel (*Methods of Investigating Highly Dispersed and Porous Materials*), Izd. Akad. Nauk SSSR, Moscow 1958, p. 85.
[66] Beebe, R. A., Kiselev, A. V., Kovaleva, N. V., Tizon, R. F. and Holmes J. M., *Zh. Fiz. Khim.*, **38**, 939 (1964).
[67] Pierotti, R. A. and Thomas, H. E., *Physical Adsorption: The Interaction of Gases with Solids*, J. Wiley, New York 1971.
[68] Steele, W. A., *The Interaction of Gases with Solid Surfaces*, Pergamon Press, Oxford 1974.
[69] Ross, S. and Olivier, J. P., *On Physical Adsorption*, Interscience–J. Wiley, New York 1964.
[70] Barker, J. A. and Everett, D. H., *Trans. Faraday Soc.*, **58**, 1080 (1962).
[71] Pierotti, R. A., *Chem. Phys. Letters*, **2**, 385 (1968).
[72] Pierotti, R. A., *Trans. Faraday Soc.*, **70**, 1725 (1974).
[73] Sams, J. R., *Progress in Surface and Membrane Science*, Vol. 8, Academic Press, New York 1974, pp. 1–47.
[74] Rudziński, W., *Phys. Letters*, **42A**, 519 (1973).
[75] Rudziński, W., *Chem. Phys. Letters*, **10**, 183 (1971).
[76] Rudziński, W., *Phys. Letters*, **58A**, 443 (1972).
[77] Rudziński, W., *Czech. J. Phys.*, **B23**, 488 (1973).
[78] Rudziński, W. and Jaroniec, M., *Surface Sci.*, **42**, 552 (1974).
[79] Toth, J., Rudziński, W., Waksmundzki, A., Jaroniec, M. and Sokołowski, S., *Acta Chem. Hung.*, **82**, 11 (1974).
[80] Halsey, G. D. and Taylor, H. S., *J. Chem. Phys.*, **15**, 624 (1947).
[81] Umeda, S. Teranishi, S. and Tarama, K., *Bull. Inst. Chem. Research, Kyoto Univ.*, **32**, 109 (1954).
[82] Rudnitski, L. A. and Alekseev, A. M., *Dokl. Akad. Nauk SSSR*, **206**, 1169 (1972); Sokołowski, S., Jaroniec, M. and Cerofolini, G. F., *Surface Sci.*, **47**, 429 (1975).
[83] Sips, R., *J. Chem. Phys.*, **16**, 490 (1948).
[84] Sips, R., *J. Chem. Phys.*, **18**, 1024 (1950).
[85] *Table of Integral Transforms*, Erdelyi, A. (Ed.), New York 1954.
[86] Misra, D. N., *J. Chem. Phys.*, **54**, 5049 (1970).
[87] Misra, D. N., *Surface Sci.*, **18**, 367 (1969).
[88] Harris, L. B., *Surface Sci.*, **10**, 129 (1968).
[89] Harris, L. B., *Surface Sci.*, **13**, 377 (1969).
[90] Harris, L. B., *Surface Sci.*, **15**, 182 (1969).
[91] Cerofolini, G. F., *Surface Sci.*, **24**, 391 (1971).
[92] Cerofolini, G. F., *J. Low. Temp. Phys.*, **6**, 471 (1972).
[93] Cerofolini, G. F., *Thin Solid Films*, **23**, 129 (1974).
[94] Adamson, A. W. and Ling, I., *Solid Surface and Gas–Solid Interface*, Interscience, New York 1961, p. 51.
[95] Dormant, L. M. and Adamson, A. W., *J. Coll. Interface Sci.*, **38**, 285 (1972).
[96] Drain, L. H. and Morrison, J. A., *Trans. Faraday Soc.*, **48**, 840 (1952).
[97] Rudziński, W., Jaroniec, M., Sokołowski, S. and Cerofolini, G. F., *Czech. J. Phys.*, **B25**, 891 (1975).
[98] Jaroniec, M. and Rudziński, W., *Colloid and Polymer Sci.*, **253**, 683 (1975).
[99] Jaroniec, M. and Rudziński, W., *Acta Chem. Hung.*, **88**, 351 (1976).
[100] Jaroniec, M. and Rudziński, W., *Surface Sci.*, **52**, 641 (1975).

[101] Van Dongen, R. H., *Surface Sci.*, **39**, 341 (1973).

[102] Rudziński, W., Waksmundzki, A., Jaroniec, M. and Sokołowski S., *Roczniki Chem.*, **48**, 1985 (1974).

[103] Jaroniec, M., Rudziński, W., Sokołowski, S. and Smarzewski, R., *Colloid and Polymer Sci.*, **253**, 164 (1975).

[104] Jaroniec, M., *Surface Sci.*, **50**, 553 (1975).

[105] Jaroniec, M., *Z. physik. Chem. (Leipzig)*, **257**, 449 (1976).

[106] Jaroniec, M., *J. Coll. Interface Sci.*, **52**, 41 (1975).

[107] Rudziński, W., Jaroniec, M. and Sokołowski, S., *Phys. Letters*, **48A**, 171 (1974).

[108] Waksmundzki, A., Rudziński, W., Sokołowski, S. and Jaroniec, M., *Roczniki Chem.*, **48**, 1741 (1974).

[109] Rudziński, W., Sokołowski, S., Jaroniec, M. and Toth, J., *Z. physik. Chem.*, **256**, 273 (1975).

[110] Jaroniec, M., Sokołowski, S. and Cerofolini, G. F., *Thin Solid Films*, **31**, 321 (1976).

[111] Brunauer, S., Skalny, J. and Bodor, E. E., *J. Coll. Interface Sci.*, **30**, 546 (1969).

[112] Misra, D. N., *J. Coll. Interface Sci.*, **43**, 85 (1973).

[113] Rudziński, W. and Jaroniec, M., *Roczniki Chem.*, **49**, 165 (1975).

[114] House, W. A. and Jaycock, M. J., *J. Coll. Interface Sci.*, **47**, 50 (1974).

[115] Waksmundzki, A., Jaroniec, M. and Łajtar, L., *Roczniki Chem.*, **49**, 1003 (1975).

[116] Waksmundzki, A., Jaroniec, M. and Suprynowicz, Z., *J. Chromatogr.*, **110**, 381 (1975).

[117] Cerofolini, G. F., *Surface Sci.*, **47**, 469 (1975).

[118] Cerofolini, G. F., *Thin Solid Films*, **26**, 53 (1975).

[119] McCafferty, E. and Zettlelmoyer, A. C., *J. Coll. Sci.*, **34**, 452 (1970).

[120] McBain, J. W. and Baker, A. M., *J. Am. Chem. Soc.*, **48**, 690 (1926).

[121] Dreving, V. P. and Eltekov, Yu. A., *Filyal Vsesoyus. Inst. Nauch. i Tekhn. Inf. Moscow*, 3 (1958).

[122] Kiselev, A. V. and Yashin, I. L., *Gazoadsorptsionnaya khromatografiya* (*Gas Adsorption Chromatography*), Izd. Nauka, Moscow 1967.

[123] Dubinin, M. M., *Quart. Rev. Chem. Soc.*, **9**, 101 (1959); *Chem. Rev.*, **60**, 235 (1960).

[124] De Boer, J. M , *The Structure and Properties of Porous Materials*, D. H., Everett and F. S. Stone (Eds.), Butterworths Scientific Publ., London 1958, p. 68.

[125] Iler, R. K., *The Colloid Chemistry of Silica and Silicates*, Cornell University Press, Ithaca, New York 1955.

[126] Linsen, B. G., *Physical and Chemical Aspects of Adsorbents and Catalysts*, B. G. Linsen (Ed.), Academic Press, London–New York 1970.

[127] Neimark, I. E., Sheinfain, P. Yu., *Silikagel, yego poluchennye, svoistva i primenenye* (*Silica Gel, its Production, Properties and Application*), Izd. Naukova Dumka, Kiev 1963.

[128] Czarny, Z., *Wiadomości Chem.*, **17**, 395 (1963).

[129] Boehm, H. P., *Angew. Chem., Int. Ed.*, **5**, 533 (1966).

[120] Unger, K., *Angew. Chem., Int. Ed.*, **11**, 267 (1972).

[131] Weyl, W. A., *Research*, **3**, 230 (1950).

[132] Hauser, E. A., *J. Phys. Coll. Chem.*, **55**, 605 (1951).

[133] Kiselev, A. V., *Poverkhnostnye khimicheskie sŏedinenya i ikh rol v yavlenyak*

adsorpcii (*Surface Chemical Compounds and their Role in Adsorption Phenomena*), Izd. MGU, Khimya, Moscow 1957, p. 90.

[134] Okkerse, C. and de Boer, J. H., *J. chim. phys.*, **57**, 534 (1960).

[135] Okkerse, C. and de Boer, J. H., *Silic. Ind.*, **27**, 195 (1962).

[136] Gärtner, K. and Griessbach, R., *Kolloid. Z.*, **160**, 21 (1958).

[137] Borezkov, G. K., Borisova, M. S., Dzhygit, O. M., Dzisko, V. A., Dreving, V. P., Kiselev, A. V. and Likhacheva, O. A., *Zh. Fiz. Khim.*, **22**, 603 (1948).

[138] Neimark, I. E. and Sheinfain, P. Yu., *Koll. Zh.*, **15**, 45, 145 (1953).

[139] Lippens, B. C., *Structure and Texture of Aluminas*, Thesis, Delft University of Technology, Delft 1961.

[140] Ginsberg, H., Hüttig, W. and Strunk-Lichtenberg, G., *Z. anorg. Chem.*, **293**, 33, 204 (1957).

[141] Krischner, H., *Ber. dt. keram. Ges.*, **43**, 479 (1966).

[142] Beretka, J. and Ridge, M. J., *J. Chem. Soc.*, *A*, 2106 (1967).

[143] Peri, J. B., *J. Phys. Chem.*, **69**, 211 (1965).

[144] Bowen, J. N., Bowrey, T. and Malin, S. H., *J. Catalysis*, **7**, 209 (1967).

[145] Johnson, M. F. L. and Mooi, J., A. C. S. Meeting, Chicago, Sept. 1967; *Petroleum Div. Reprints*, **12** (3) 205 (1967).

[146] Dubinin, M. M., *Usp. Khim.*, **24**, 3 (1955)

[147] Meier, W. M., *Zeolite Structures*, in: "*Molecular Sieves*", *Soc. Chem. Ind.*, London 1968, p. 10.

[148] Dzhygit, O. M., Kiselev, A. V. and Neimark, I. E., *Zh. Fiz. Khim.*, **28**, 1804 (1954).

[149] Temkin, M. I. and Pyzhev, M. V., *Acta Phys. URSS*, **12**, 327 (1940).

[150] Tronell, B. M., *Chemisorption*, Academic Press, New York 1955.

[151] Trykorn, F. G. and Wyatt, W. F., *Trans. Faraday Soc.*, **22**, 135 (1926).

Chapter 4

Adsorption at the Solid/Liquid Interface

4.1 Introduction

Adsorption from solutions differs from adsorption of individual substances (gases, vapours and pure liquids) in that the solution contains at least two components which can form a compact layer on the adsorbent surface. A change in concentration of the component of the solution results in their mutual displacement from the surface layer, the most characteristic feature of adsorption from solution. Since there are no vacancies in the surface solution (surface layer) and in the bulk solution, only substitution of the molecules of one component by those of the other is possible.

Of course, for the surface layer we use the Guggenheim and Adam equation (2.21).

The fundamental definitions of surface excesses described previously are also valid for adsorption at the solid/liquid (solution) interface. Adsorption in such a system (component i concentration profile) is presented schematically in Fig. 3.1, where the gas phase should now be replaced by the liquid (solution) phase. Furthermore, possible concentration changes in the surface layer on the solid (adsorbent) side are neglected.

Reduced adsorption $n_i^{\sigma(n)}$ is the excess number of moles of component i in the system as compared with the number of moles of that component in the reference system (where adsorption does not occur) containing the same total number of moles of all components in the liquid phase, $n=\sum_i n_i$, and the same concentration of component i in that phase. We can thus write:

$$n_i^{\sigma(n)}=n_i-nx_i \tag{4.1}$$

where n_i is the total number of moles of component i in the system, and x_i is the mole fraction of this component in the solution in equilibrium with the adsorbent (i.e. in the so-called equilibrium solution). The corresponding numbers of moles n, n_i and $n_i^{\sigma(n)}$ are frequently referred to 1 g of adsorbent, so that $\Gamma_i^{(n)} = n_i^{\sigma(n)}/s$ (where s is the specific surface area of the adsorbent). By definition (cf. Chapter 2), the total reduced surface excess amount of the solution components equals zero, i.e.:

$$\sum_i n_i^{\sigma(n)} = 0 \text{ and } \sum_i \Gamma_i^{(n)} = 0 . \tag{4.2}$$

The excess $n_i^{\sigma(v)}$ can be written:

$$n_i^{\sigma(v)} = n_i - c_i V^{\mathrm{l}} \tag{4.3}$$

where $V^{\mathrm{l}} = V - V_{\mathrm{a}}$ is the volume of the liquid phase (V being the volume of the system, and V_{a} being the volume of the solid phase, i.e. adsorbent), $n_i^{\sigma(v)}$ is the excess number of moles of component i in the system compared with the number of moles of that component in the reference system containing the same volume of solution V^{l} of the same concentration c_i. The volume V^{l} and excess amount $n_i^{\sigma(v)}$ are also referred to 1 g of adsorbent, hence $\Gamma_i^{(v)} = n_i^{\sigma(v)}/s$. Assuming that the partial molar volumes V_i of the components of the solution are independent of concentration and of adsorption we can write:

$$\sum_i V_{\mathrm{m},i} n_i^{\sigma(v)} = 0 \text{ and } \sum_i V_{\mathrm{m},i} \Gamma_i^{(v)} = 0 . \tag{4.4}$$

The total number of moles of component i in the surface layer can be expressed by the equation:

$$n_i^{\mathrm{s}(n)} = n_i^{\sigma(n)} + x_i n^{\mathrm{s}} = n_i^{\sigma(n)} + x_i \sum_i n_i^{\mathrm{s}} \tag{4.5}$$

or

$$n_i^{\mathrm{s}(v)} = n_i^{\sigma(v)} + c_i V^{\mathrm{s}} = n_i^{\sigma(v)} + c_i \sum_i n_i^{\mathrm{s}} V_{\mathrm{m},i}^{\mathrm{s}} \tag{4.6}$$

where $n^{\mathrm{s}} = \sum_i n_i^{\mathrm{s}}$ is the total number of moles in the surface layer, V^{s} is the volume of that layer, x_i is the mole fraction and c_i the concentration (mole/ /dm^3) of component i in the equilibrium solution. The values n_i^{s} can be estimated only when the thickness of the surface layer l^{s} and its location in the system are determined. Only in the case of an ideal adsorption layer and ideal bulk equilibrium solution do we have $n_i^{\mathrm{s}(n)} = n_i^{\mathrm{s}(v)} = n_i^{\mathrm{s}}$.

4.2 Adsorption from Non-electrolyte and Weak Electrolyte Solutions

4.2.1 Adsorption from Binary Liquid Mixtures

The Freundlich isotherm (3.22) as well as the Langmuir (3.27) or (3.29) isotherms have often been used to describe adsorption from dilute binary solutions. We replace the pressure p by concentration (x or c), and these equations adequately describe the adsorption of strongly adsorbed substances.

Note. Attempts have also been made to apply the Polanyi potential theory to adsorption from solutions [1–3].

A number of theories have been developed for adsorption isotherms describing adsorption from solutions over the whole concentration range. This problem has not yet been fully solved. Detailed studies have been carried out in the last two decades, by Kiselev [4–7], Everett [8, 9], Kipling [10], Schay [11–13], Laryonov [14–17], Šiškova and Erdös [18–20], and Rusanov [21, 22].

Let us now consider in more detail adsorption on a homogeneous adsorbent surface from a binary liquid mixture whose components 1 and 2 are completely miscible. We can then assume that at adsorption equilibrium we have two binary solutions (two phases): the surface solution (surface phase or surface layer) and the bulk solution. At thermodynamic equilibrium:

$$\mu_1^s = \mu_1, \qquad \mu_2^s = \mu_2 \tag{4.7}$$

where μ_1^s and μ_2^s are the chemical potentials of the components 1 and 2, respectively, in the surface solution, and μ_1 and μ_2 are the corresponding potentials in the bulk.

Adopting the simplest of the thermodynamic descriptions of adsorption equilibria used initially by Fu, Hansen and Bartell [23] and subsequently also by other authors [24–27] we can rewrite equilibrium (4.7) in the form:

$$\mu_1^{\ominus,s} = RT \ln x_1^s f_1^s = \mu_1^{\ominus} + RT \ln x_1 f_1 \tag{4.8}$$

$$\mu_2^{\ominus,s} + RT \ln x_2^s f_2^s = \mu_2^{\ominus} + RT \ln x_2 f_2 \tag{4.9}$$

where $\mu_1^{\ominus,s}$ and $\mu_2^{\ominus,s}$ are the standard chemical potentials of component 1 and 2 in the surface solution, $\mu_1^{\ominus}$ and $\mu_2^{\ominus}$ are those in the bulk solution, x_i^s and x_i ($x_i^s = n_i^s/n^s$), and f_i^s and f_i are accordingly the mole fractions and the activity coefficients of the respective components in the surface and bulk solutions. Of course the standard chemical potentials of the components depend on the reference system used (e.g. symmetric or asymmetric [28]).

From equations (4.8) and (4.9) we get:

$$\frac{x_1^s}{x_1}=\frac{f_1}{f_1^s}\exp\left[-\frac{1}{RT}(\mu_1^{\ominus,s}-\mu_1^{\ominus})\right] \tag{4.10}$$

and

$$\frac{x_2^s}{x_2}=\frac{f_2}{f_2^s}\exp\left[-\frac{1}{RT}(\mu_2^{\ominus,s}-\mu_2^{\ominus})\right]. \tag{4.11}$$

Dividing equation (4.10) by (4.11), side by side, we get:

$$\frac{x_1^s x_2}{x_1 x_2^s}=\frac{f_1 f_2}{f_1^s f_2}\exp\left\{-\frac{1}{RT}[(\mu_1^{\ominus,s}-\mu_1)-(\mu_2^{\ominus,s}-\mu_2^{\ominus})]\right\}. \tag{4.12}$$

Let us denote:

$$\alpha=\frac{x_1^s x_2}{x_1 x_2^s}, \tag{4.13}$$

$$C=\frac{f_1 f_2^s}{f_1^s f_2} \tag{4.14}$$

and

$$K_1=\exp\left\{-\frac{1}{RT}[(\mu_1^{\ominus,s}-\mu_1^{\ominus})-(\mu_2^{\ominus,s}-\mu_2^{\ominus})]\right\}. \tag{4.15}$$

NOTE. The difference $\mu_i^{\ominus,s}-\mu_i^{\ominus}=\Delta_a\mu_i^{\ominus}$ may be used as a measure of the adsorbability of substance i, i.e. of its adsorption affinity in the given adsorbent–solvent system [25].

Using the notations (4.13), (4.14) and (4.15) we can write equation (4.12) in the form:

$$\frac{x_1^s x_2}{x_1 x_2^s}=CK_1=\alpha. \tag{4.16}$$

Since $x_2^s=1-x_1^s$ and $x_2=1-x_1$, we have:

$$x_1^s=\frac{\alpha x_1}{\alpha x_1+x_2}=\frac{\alpha x_1}{1+(\alpha-1)x_1}. \tag{4.17}$$

This is the general adsorption isotherm for binary solutions, representing the variation of surface solution composition (x_1^s) with bulk composition (x_1), and therefore it is often called the individual adsorption isotherm of component 1 from solution. In equation (4.17) the quantity α (also denoted by f or K), defined by equation (4.13), is known as the distribution coefficient or distribution function. This depends on the composition of the bulk solution, and therefore the general adsorption isotherm for a binary solution (4.17) is not an adsorption equilibrium equation.

For a uniform adsorbent surface, $K_1=\text{const}$. When adsorption takes place from an ideal solution and the surface solution is also ideal, then $C=1$, and:

$$\frac{x_1^s x_2}{x_1 x_2^s}=K_1 \tag{4.18}$$

where K_1 is the adsorption equilibrium constant. Hence:

$$x_1^s=\frac{K_1 x_1}{1+(K_1-1)x_1}\,. \tag{4.19}$$

Equation (4.19) has also been derived by Everett by statistical methods [8] for the case of formation of an ideal adsorption monolayer. This equation, after suitable transformation, is widely used in studies of adsorption from solutions (see for instance Section 4.2.3).

According to Butler [29], Everett [9], and Rusanov [21] thermodynamic adsorption equilibrium in the solid–solution system can be described thermodynamically in an alternative way. The Holmholtz free energy of the surface layer at $T=\text{const}$ can be written in the form (cf. equation (2.32)):

$$dF^s=\gamma\, dA+\sum_i \mu_i dn_i^s \tag{4.20}$$

where γ is the interfacial tension at the solid/solution interface, and A is the surface area of the adsorbent. Then:

$$\mu_i^s=\left(\frac{\partial F^s}{\partial n_i^s}\right)_{T,A,V^s,n_j^s}=\gamma\left(\frac{\partial A}{\partial n_i^s}\right)_{T,V^s,\gamma,n_i^s}+\mu_i\,. \tag{4.21}$$

If we put

$$\left(\frac{\partial A}{\partial n_i^s}\right)_{T,V^s,\gamma,n_j^s}=a_i \tag{4.22}$$

then, after taking the product γa_i over to the left side of equation (4.21), we write:

$$\mu_i^{\ominus,s}+RT\ln x_i^s f_i^s-\gamma a_i=\mu_i^{\ominus}+RT\ln x_i f_i\,. \tag{4.23}$$

If we assume the pure components of a binary solution 1+2 to be the reference system, we can write the adsorption equilibrium as:

$$\mu_1^{0,s}+RT\ln x_1^s f_1^s+(\gamma-\gamma_1^0)\,a_1=\mu_1^0+RT\ln x_1 f_1 \tag{4.24}$$

$$\mu_2^{0,s}+RT\ln x_2^s f_2^s+(\gamma-\gamma_2^0)\,a_2=\mu_2^0+RT\ln x_2 f_2 \tag{4.25}$$

where γ_1^0 and γ_2^0 are tensions between the phases at the pure component 1/adsorbent and pure component 2/adsorbent interfaces. Assuming that the chemical potentials of the pure substances 1 and 2 are equal in the surface layer and in the bulk phase, i.e. that $\mu_1^{0,s}=\mu_1^0$ and $\mu_2^{0,s}=\mu_2^0$, we find (after similar transformations as in the case of equations (4.10) and (4.11):

$$\frac{x_1^s x_2}{x_1 x_2^s}=C\exp\left\{-\frac{1}{RT}\left[(\gamma_1^0-\gamma)a_1-(\gamma_2^0-\gamma)a_2\right]\right\} \tag{4.26}$$

or

$$K_1=\exp\left\{-\frac{1}{RT}\left[(\gamma_1^0-\gamma)a_1-(\gamma_2^0-\gamma)a_2\right]\right\}. \tag{4.27}$$

When $a_1=a_2=a$, then

$$K_1=\exp\left[-\frac{1}{RT}(\gamma_1^0-\gamma_2^0)a\right]. \tag{4.28}$$

If we now compare the left-hand sides of equations (4.8) and (4.9), and equations (4.24) and (4.25) we find that the formulations of surface layer chemical potentials are different for the two thermodynamic descriptions of adsorption equilibrium. In the first approach, when the pure substance i is the reference system, we write:

$$\mu_i^s=\mu_i^{0,s}+RT\ln x_1^s f_i^{s,*}. \tag{4.29}$$

In terms of the approach of Butler and others we find:

$$\mu_i^s=\mu_i^{0,s}+RT\ln x_i^s f_i^s-(\gamma-\gamma_i^0)a_i. \tag{4.30}$$

We can then derive the following relation between the activity coefficients of substance i in the surface layer used in the two thermodynamic descriptions of adsorption equilibrium:

$$f_i^{s,*}=f_i^s\exp\left[-(\gamma-\gamma_i^0)a_i/RT\right]. \tag{4.31}$$

Since $\gamma\neq\gamma_i^0$, the quantity $f_i^{s,*}$ will not equal unity in the ideal system. On the other hand it is not possible to measure γ and γ_i^0. It seems that the use of the first thermodynamic description of the adsorption equilibrium is justified when studying some adsorption problems.

In case of adsorption from a non-ideal solution the value of C varies with mole fraction x_1, and therefore α also varies with x_1.

For every value of α, when $x_1\to 0$ (beginning of the adsorption isotherm for component 1), $1+(\alpha-1)\,x_1\to 1$ and then:

$$x_1^s=\alpha x_1. \tag{4.32}$$

When $x_1=1$ (end of the isotherm), then

$$x_1^s=x_1=1 \tag{4.33}$$

which means that the surface layer consists of pure component 1.

Let us now consider the effect of the magnitude of C and K_1 on the value of α. If component 1 is strongly adsorbed, i.e. when $-(\mu_1^{\ominus,s}-\mu_1^{\ominus})$

is very large and greater than RT, then $K_1 \gg 1$. In that case we also have $\alpha \gg 1$. Slight deviations of C from unity cannot affect this argument, i.e. $\alpha \gg 1$ for all values of x_1, and in the adsorption isotherm (4.17) $(\alpha - 1) \approx \alpha$. Hence we can write:

$$x_1^s \approx \frac{\alpha x_1}{1 + \alpha x_1}. \tag{4.34}$$

This latter equation is similar to the Langmuir equation.

If $\alpha \gg 1$, then, for small values of x_1, $\alpha_1 x_1 \gg 1$ and x_1^s is close to unity over a wide range of concentration. In Fig. 4.1 this is represented by curve *1* which shows that component 1 accumulates in the surface solution.

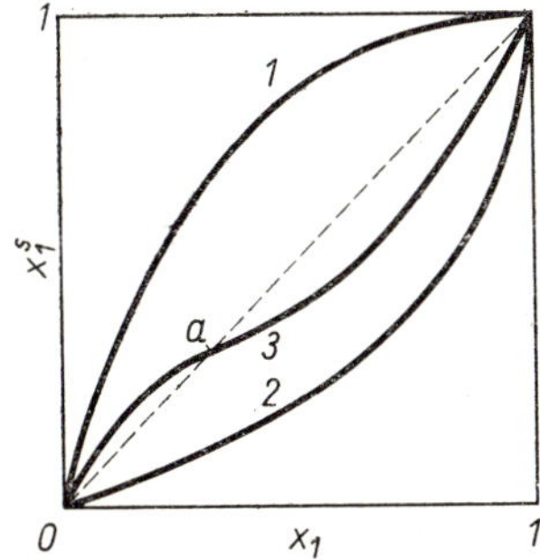

FIG. 4.1 Individual adsorption isotherms from a binary solution (the concentrations of the given components are in mole fractions): *1*—positive adsorption of component 1 over the whole range of concentrations, *2*—negative adsorption of component 1 over the whole range of concentrations, *3*—limited adsorption of both components of the solution, *a*—the adsorption azeotropic point.

If $K_1 \ll 1$, then, for small deviations of C from unity, we have $\alpha \ll 1$ for all values of x_1. In that case we have in equation (4.17) $\alpha - 1 \approx -1$, and so, for small values of x_1, we get:

$$x_1^s \approx \frac{\alpha x_1}{1 - x_1} \tag{4.35}$$

i.e. the mole fraction, x_1^s, in the surface solution remains small over a wide range of x_1. Only when x_1 approaches unity does x_1^s begin to increase rapidly, as represented by curve *2* in Fig. 4.1.

If $K_i \approx 1$, i.e. when the adsorbabilities of the components of a binary solution do not differ significantly from each other, then the value of α

is affected both by the heterogeneity of the surface (K_1 varies with increasing x_1^s and x_1) and the deviation of the surface and bulk solutions from ideality, i.e. deviations of the value of C from unity. In such cases the value of α is greater than unity for small values of x_1, and smaller than unity for larger values of x_1 (sometimes the opposite may be true). The difference $\alpha-1$ changes its sign with increasing x_1; this is represented by curve *3* in Fig. 4.1 which intersects the straight line $x_1^s=x_1$. At the point of intersection *a* the composition of the surface solution and that of the bulk solution are equal. This phenomenon has been termed by Schay [13] adsorption azeotropy.

Curves *1*, *2* and *3* in Fig. 4.1 represent all possible types of adsorption isotherms for a binary solution. It should be emphasized that we consider here the total amount of given substance (expressed in terms of the mole fraction) in the surface layer, and x^s is often equal to the surface coverage θ (in the case of monolayer adsorption).

Denoting the numbers of moles of the components in the surface solution on 1 g of adsorbent at adsorption equilibrium by n_1^s and n_2^s, we can write:

$$n_i^{\sigma(n)}=n_1^s-(n_1^s+n_2^s)x_1 \tag{4.36}$$

(this equation is derived later in this chapter).

If we denote the maximum number of moles of the components which can be present in the surface layer of 1 g of adsorbent by $n_{m,1}^s$ and $n_{m,2}^s$, we can write:

$$\frac{n_1^s}{n_{m,1}^s}+\frac{n_2^s}{n_{m,2}^s}=1. \tag{4.37}$$

Then:

$$n_2^s=n_{m,2}^s-\beta n_1^s \tag{4.38}$$

where $\beta=n_{m,2}^s/n_{m,1}^s$ is the coefficient of surface displacement of components 1 and 2. We can also write that $\beta=\omega_{m,1}/\omega_{m,2}$, where $\omega_{m,1}$ and $\omega_{m,2}$ are the adsorbent surface areas of the adsorption layer occupied by the molecules of the solution components. On the other hand, from the definition of mole fraction:

$$x_1^s=\frac{n_1^s}{n_1^s+n_2^s}. \tag{4.39}$$

If we substitute the expression for n_2^s from equation (4.38) into equation (4.39) and introduce the resultant expression obtained for x_1^s into equation

(4.17) and solve the latter with respect to n_1^s, we obtain the adsorption isotherm for binary solutions in the form:

$$n_1^s = \frac{n_{m,2}^s \alpha x_1}{1+(\beta\alpha-1)x_1} \tag{4.40}$$

Since $\beta = n_{m,2}^s / n_{m,1}^s$, we can write:

$$n_{m,2}^s = \beta n_{m,1}^s \,. \tag{4.41}$$

Taking into account equation (4.41), equation (4.40) becomes:

$$n_1^s = \frac{n_{m,1}^s \beta\alpha x_1}{1+(\beta\alpha-1)x_1} \,. \tag{4.42}$$

The adsorption isotherms described by this equation are shown in Fig. 4.2. The numbers of moles n_1^s and n_2^s of the solution components in the surface layer of 1 g of adsorbent are often referred to as their real or individual adsorption.

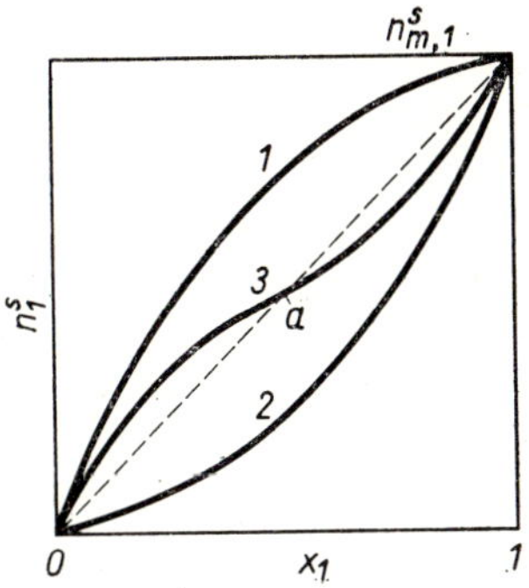

FIG. 4.2 Individual adsorption isotherms from a binary solution (the quantities of the given component are in moles per gram of adsorbent): *1* – positive adsorption of component 1 over the whole range of concentrations, *2* – negative adsorption of component 1 over whole range of concentrations, *3* – limited adsorption of both components of the solution, *a* – the adsorption azeotropic point.

Substituting the expression for n_2^s from equation (4.38) into equation (4.36) and for n_1^s the expression (4.42) we obtain the excess adsorption (reduced adsorption) isotherm $n_1^{\sigma(n)} = f(x_1)$ in the form:

$$n_1^{\sigma(n)} = \frac{n_{m,1}^s \beta(\alpha-1)x_1(1-x_1)}{1+(\beta\alpha-1)x_1} \,. \tag{4.43}$$

If we let $\alpha^{\mathrm{s}}_{\mathrm{m},1} = n^{\mathrm{s}}_{\mathrm{m},1}/s$ and take account of equation (4.43), the adsorption isotherm can be written as:

$$\Gamma_1^{(n)} = \frac{\alpha^{\mathrm{s}}_{\mathrm{m},1}(\alpha-1)x_1(1-x_1)}{1+(\beta\alpha-1)x_1}. \tag{4.44}$$

The quantity $\Gamma_1^{(n)}=0$ for any value of α when $x_1=0$ or $x_1=1$; the same is true for $n_1^{\sigma(n)}$. If $\alpha=1$, then for any value of x_1 we have $n_1^{\sigma(n)}=0$ and $\Gamma_1^{(n)}=0$. For $\alpha \gg 1$, the isotherm $\Gamma_1^{(n)}=\varphi(x_1)$ initially rises rapidly until it reaches its maximum and subsequently falls practically linearly to zero, since for larger x_1 values and $\alpha \gg 1$ equation (4.44) becomes linear:

$$\Gamma_1^{(n)} \approx \alpha^{\mathrm{s}}_{\mathrm{m},1}(1-x_1) \tag{4.45}$$

and similarly,

$$n_1^{\sigma(n)} \approx n^{\mathrm{s}}_{\mathrm{m},1}(1-x_1). \tag{4.46}$$

When $\alpha \approx 1$, then the shape of the adsorption isotherm is determined by the value and sign of the difference $\alpha-1$, i.e. the departure of the surface from homogeneity and of the bulk and surface solution from ideality. When the difference $\alpha-1$ changes sign, then $\Gamma_1^{(n)}$ also changes sign, assuming the value of zero at the point of adsorption azeotropy. All the above cases of excess adsorption isotherms are presented in Fig. 4.3.

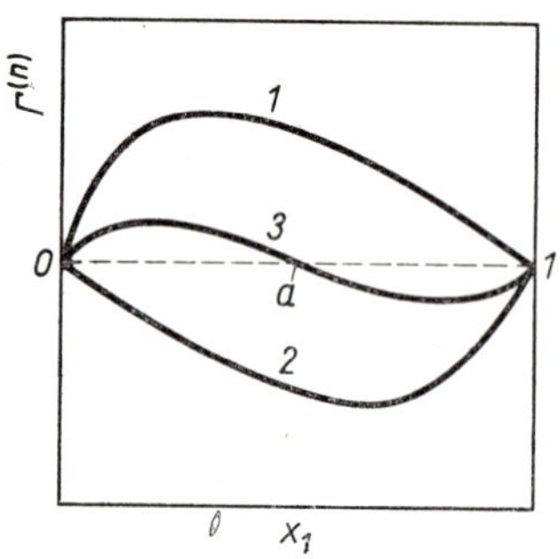

Fig. 4.3 Excess adsorption isotherms for a binary solution: *1*—positive adsorption of component 1 over the whole range of concentrations, *2*—negative adsorption of component 1 over the whole range of concentrations, *3*—limited adsorption of both components of the solution, *a*— the adsorption azeotropic point.

It is seen that the adsorption isotherms for binary solutions can be represented by the three corresponding equations: (4.17), (4.42) and (4.43) or (4.44), depending on how adsorption is expressed. All adsorption isotherms include the important quantity α which determines the shape of the isotherm, the presence of one or several extremes and their location as well as that of the points of adsorption azeotropy. From the expression determining the magnitude of $\alpha = CK_1$ we see that adsorption of the given component from solution depends not only on the interaction of adsorbate molecules with the adsorbent, as is the case for gas adsorption at low pressures, but on the resultant of their interactions with the adsorbent (K_1) and with the bulk and surface solutions (C). Adsorption from solution is therefore generally much smaller than that on the solid/gas (vapour) interface.

Taking account of equations (4.36), (4.37) and (4.39), and using the notation $\beta = n^s_{m,2}/n^s_{m,1}$ we get the following expressions, often useful for calculations [30]:

$$n^s_1 = \frac{n^s_{m,1}\beta x_1 + n_1^{\sigma(n)}}{1+(\beta-1)x_1}, \tag{4.47}$$

$$x^s_1 = \frac{n^s_{m,1}\beta x_1 + n_1^{\sigma(n)}}{n^s_{m,1}\beta - (\beta-1)n_1^{\sigma(n)}} \tag{4.48}$$

we also get:

$$\alpha = \frac{(n^s_{m,1}\beta x_1 + n_1^{\sigma(n)})(1-x_1)}{\beta[n^s_{m,1}(1-x_1) - n_1^{\sigma(n)}]x_1}. \tag{4.49}$$

It is useful to classify the excess adsorption isotherms. The first classification was made by Ostwald and Izaguirre [31] and was based on isotherm shape, as shown in Fig. 4.4. The first two types of isotherms are quite common and will be discussed later. Linear isotherms correspond to the situation when molecular sieves are used as adsorbents, i.e. only one component of the solution is retained.

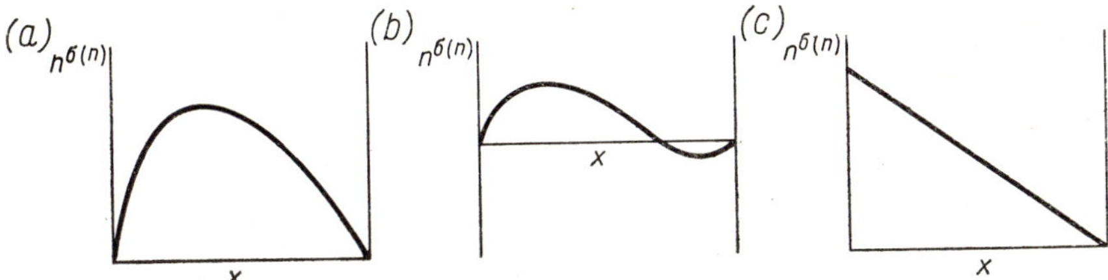

FIG. 4.4 Classification of excess adsorption isotherms for a binary solution (after Ostwald and Izaguirre [31]): (a) U-shaped isotherm consisting of one part, (b) S-shaped isotherm consisting of two parts: that for positive and that for negative adsorption, (c) linear isotherm.

A more detailed classification of the first two types of surface excess isotherms for adsorption from solutions was proposed by Schay and Nagy [11, 32] as late as 1960 (Fig. 4.5).

The excess adsorption isotherms allow us to detect the presence and estimate the location of the extremes of these isotherms. For instance, the position of the excess adsorption $n_1^{\sigma(n)}$ extreme can be found from the analysis of the extreme of function (4.43). Assuming that the quantities α and β

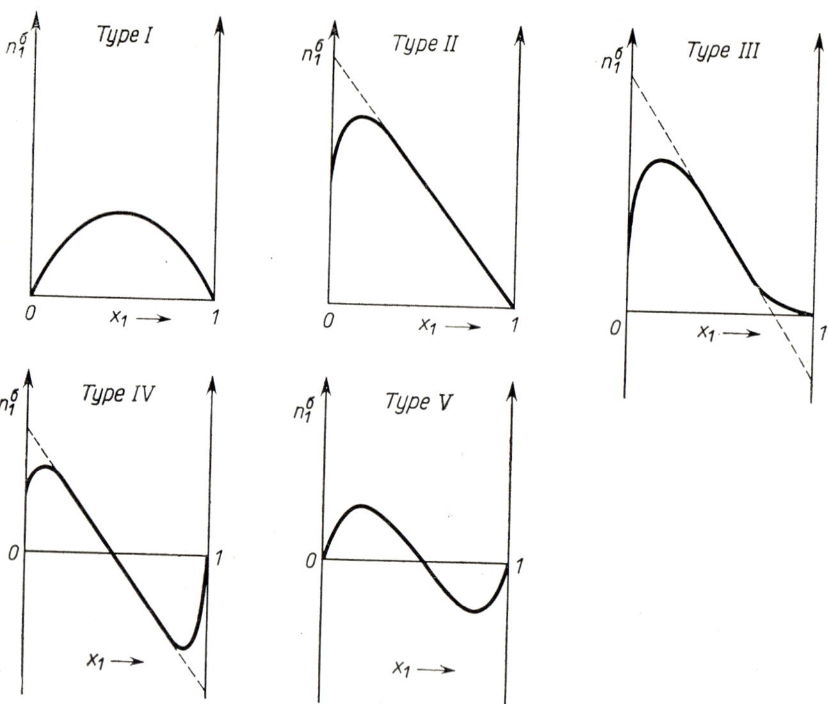

FIG. 4.5 Classification of excess adsorption isotherms for a binary solution (after Nagy and Schay [11, 32]).

in this equation are independent of mole fraction x_1, we obtain the mole fraction corresponding to the extrema from the formula:

$$x_{1,\text{ex}} = \frac{1}{1+(\alpha\beta)^{\frac{1}{2}}}. \tag{4.50}$$

For substances which are strongly adsorbed ($\alpha\beta \gg 1$), we can write equation (4.50) in the form:

$$x_{1,\text{ex}} \approx \frac{1}{(\alpha\beta)^{\frac{1}{2}}} = \left(\frac{n_{\text{m},1}^{\text{s}}}{\alpha n_{\text{m},2}^{\text{s}}}\right)^{\frac{1}{2}}. \tag{4.51}$$

It follows that with increasing α, i.e. with increasing adsorption energy and decreasing $n^s_{m,1}$ (related to the increase of size of the adsorbed molecules of component 1), the excess adsorption isotherm maximum is shifted towards smaller equilibrium concentrations of the bulk solution. The course of excess adsorption isotherms, i.e. relation of $x^s_1 - x_1 = (1/n^s)n_1^{\sigma(n)}$ (see equation (4.64)) to x_1 at various positive values of K_1 is shown in Fig. 4.6, which is based on the paper by Everett [8] and on isotherm (4.19). It has been assumed that the surface and bulk solutions are ideal (n^s being the sum of n^s_1 and n^s_2).

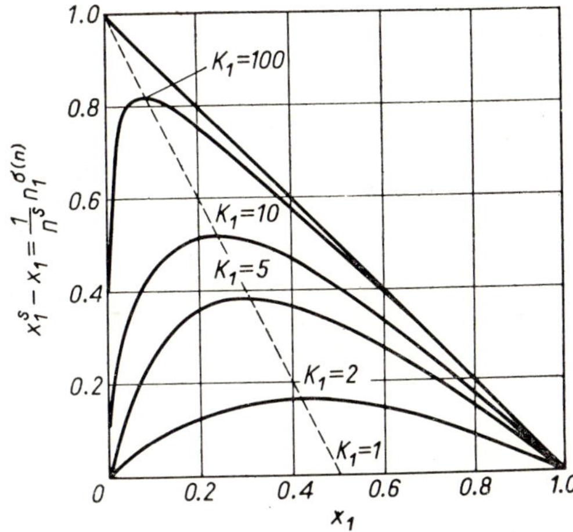

FIG. 4.6 Excess adsorption isotherms for ideal binary solutions at different positive K_1 values (after Everett [8]).

Schay [33] and subsequently Everett [9] have discussed adsorption from binary regular solutions. Assuming a lattice model for the solution and the formation of a regular adsorption monolayer, Everett obtained the following equation describing adsorption equilibrium:

$$\ln \frac{x^s_1 x_2}{x_1 x^s_2} = \ln K_1 + (l+m)\frac{v}{RT}(x_2 - x_1) + 1\frac{v}{RT}(x^s_1 - x^s_2) \qquad (4.52)$$

where K_1 is the adsorption equilibrium constant (defined by equation (4.18)) for an ideal solution and adsorption layer, l and m are quantities related to the lattice model of the liquid, and v is a constant dependent on the interaction energy between molecules of the solution components.

Equation (4.52) can be solved and we obtain x_1^s as a function of x_1 for given values of K_1 and v. For the most tightly filled spatial lattice we

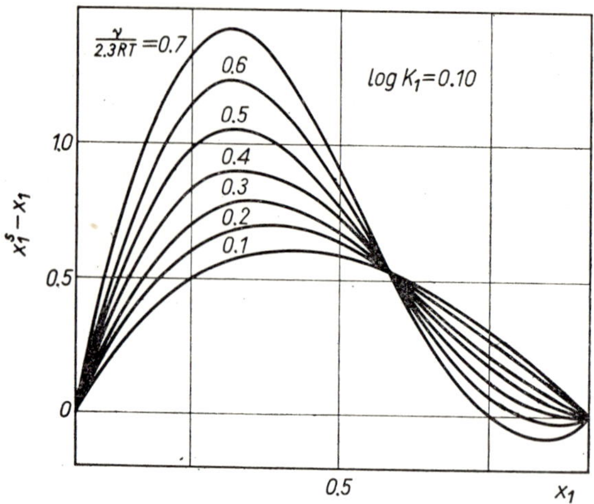

FIG. 4.7 Excess adsorption isotherms for regular binary solutions as calculated from equation (4.52) for log K_1=0.10 and $v/2.3RT$=0.1 to 0.7 (after Everett [9]).

have $l=1/2$ and $m=1/4$. In Fig. 4.7 the function

$$x_1^s - x_1 = \frac{n_1^{\sigma(n)}}{n^s} = f(x_1)$$

is plotted assuming log K_1=0.10 and $v/2.3RT$ lying in the range of 0.1–0.7, whereas Fig. 4.8 presents the plot of the function

$$x_1^s - x_1 = \frac{n_1^{\sigma(n)}}{n^s} = f(x_1)$$

assuming that $v/2.3RT$=0.50 and log K_1 varies from 0 to 0.30.

For a given value v we observe adsorption azeotropy in a limited range of values of K_1:

$$\frac{-mv}{RT} \leqslant \ln K_1 \leqslant \frac{mv}{RT}. \tag{4.53}$$

The value of v is limited in regular solutions by the value $2RT$, at which the so-called phase separation occurs. Thus in that case the phenomenon of adsorption azeotropy is limited to mixtures for which K_1 does not depart much from unity.

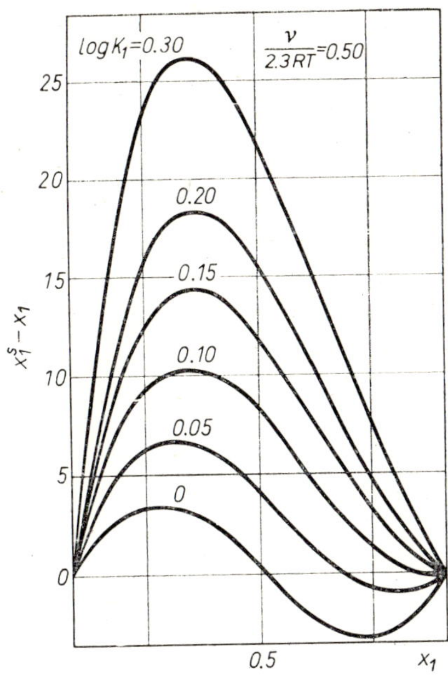

Fig. 4.8 Excess adsorption isotherms for regular binary solutions as calculated from equation (4.52) for $\nu/2.3RT=0.5$ and log K_1 values ranging from 0 to 0.30 (after Everett [9]).

4.2.2 Adsorption from Binary Solutions of Substances of Limited Miscibility

The adsorption isotherms of binary solutions of liquids of limited miscibility often have the shape shown in Fig. 4.9, which illustrates adsorption on silica gel of methyl alcohol from heptane [34]. Such isotherms are referred to as S-shaped but we should distinguish them, however, from S-shaped isotherms obtained in the case of adsorption from a binary mixture of unlimited miscibility. We see from Fig. 4.9 that adsorption increases rapidly as its solubility limit is approached (the abscissa shows the relative or reduced concentration c/c_n, i.e. the ratio of the equilibrium concentration c and that of the saturated solution, c_n) and tends asymptotically to a line parallel to the adsorption (ordinate) axis.

It is very difficult to determine such an isotherm at concentrations close to saturation. The rapid increase of adsorption at these concentrations suggests that we are dealing with a process of multilayer adsorption. More recent studies have shown that the rapid rise of the isotherm in its final section is due to phase separation which starts earlier than in the bulk because of the effect of the porous structure of the adsorbent. This process is therefore similar to the capillary condensation observed in vapour adsorption.

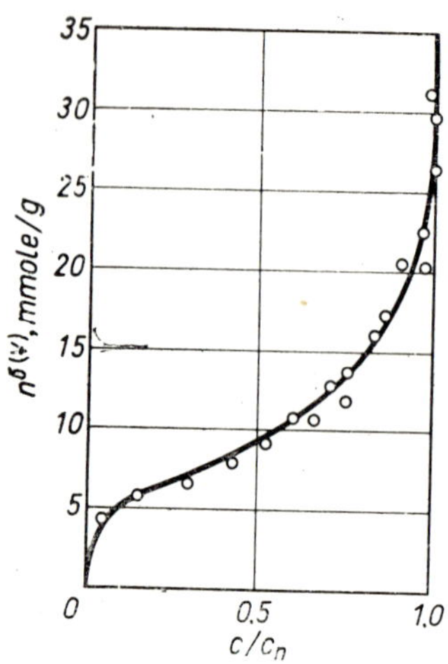

FIG. 4.9 Excess adsorption isotherm of methanol on silica gel from *n*-heptane (after Dzhygit *et al.* [34]).

Adsorption can be studied at temperatures exceeding the critical dissolution temperature (upper and lower) when the solubility of the liquid in the solution becomes unlimited. This can be described, based on work of Kiselev and Kulichenko [35], in terms of adsorption of triethylamine from aqueous solutions. The lower consolute temperature for triethylamine in water is 291 K. Its adsorption isotherm from aqueous solution on wide-porous activated carbon is shown in Fig. 4.10 for two temperatures: one above the consolution point (305.7 K) and one below that point (273 K). Below the consolution temperature the shape of the triethylamine adsorption isotherm corresponds to that of the isotherms of completely miscible liquids (this is, of course, the beginning of the curve).

The most common and typical shape of adsorption isotherm for adsorption of a solid from solution is illustrated in Fig. 4.11. Only rarely has the adsorption isotherm of a solid from solution been found to behave differently.

A solid liquefies at a sufficiently high temperature. Therefore, if we consider adsorption at two temperatures, higher and lower than the freezing

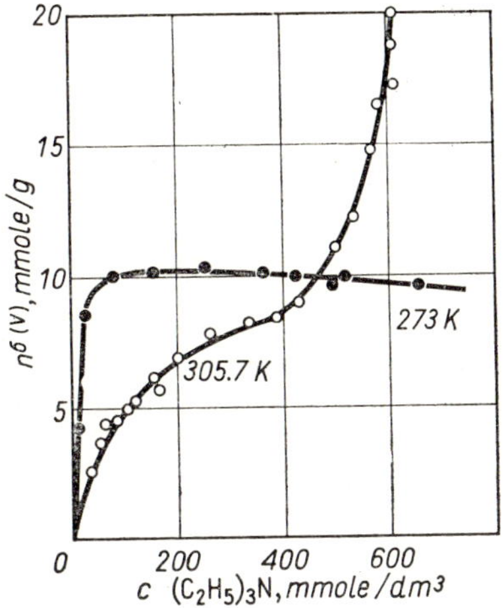

FIG. 4.10 Excess adsorption isotherms of triethylamine on wide-porous activated carbon from aqueous solutions at 305.7 K and 273 K (after Kiselev and Kulichenko [35]).

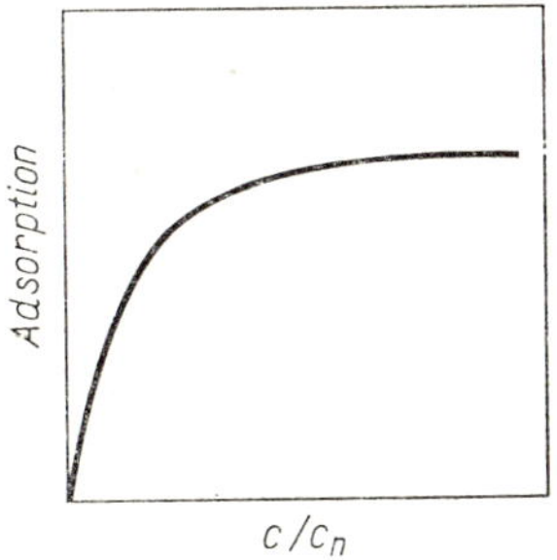

FIG. 4.11 Typical adsorption isotherm of a solid from a binary solution.

point of the substance being adsorbed, we shall obtain two isotherms:

(i) typical for adsorption of a solid from solution,

(ii) S-shaped, typical for adsorption of a liquid of partial solubility.

Figure 4.12 illustrates isotherms of phenol adsorption from *n*-heptane on a wide-porous silica gel at 293 and 313 K, after Krasilnikov and Kiselev [36].

Adsorption isotherms of solids from solution have been classified by Giles *et al.* [37]. Four principal classes of adsorption isotherms are distin-

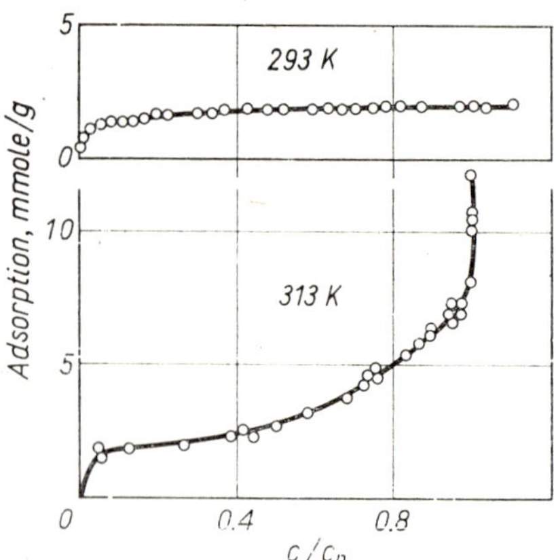

FIG. 4.12 Excess adsorption isotherms of phenol from *n*-heptane on wide-porous silica gel at 293 and 313 K (after Krasilnikow and Kiselev [36]).

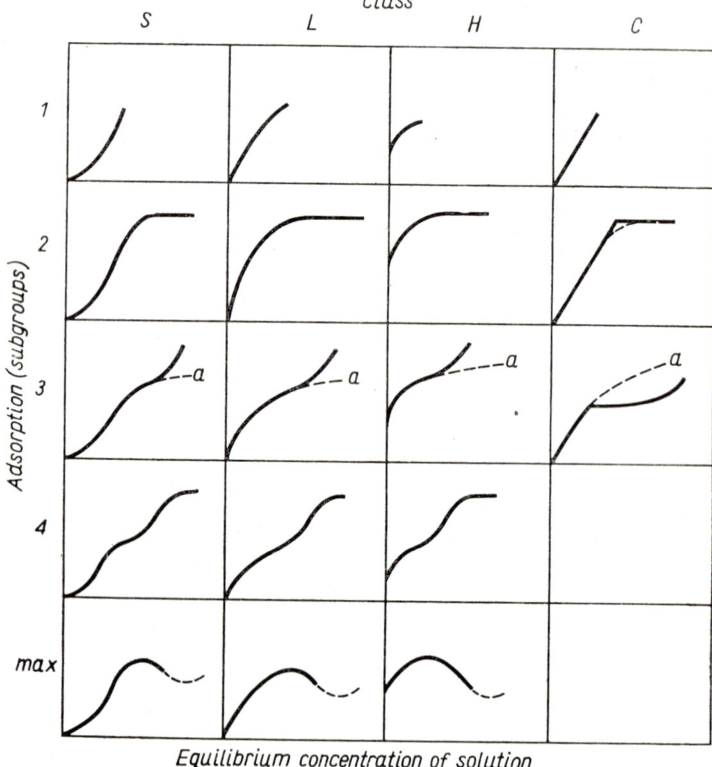

FIG. 4.13 Classification of adsorption isotherms of solids from solution (after Giles *et al.* [37]).

guished in terms of their initial sectors (Fig. 4.13): class *S*, class *L* (of Langmuir type), class *H* (high adsorption affinity), and class *C* (constant partition of the substance between the surface layer and the bulk phase). Within the particular classes, subgroups 1, 2, 3, 4 and max (revealing a maximum) are distinguished by the shape of the isotherms at higher concentration. Giles *et al.* [38, 39] relate the adsorption mechanism, orientation of adsorbed molecules and orientation changes, etc. to the shape of the isotherm. In the case of class *S* isotherms the orientation of the adsorbate molecules in the surface layer is vertical or possibly inclined, and in the case of class *L* isotherms the orientation of the adsorbate molecules on the adsorbent surface is horizontal (e.g. flat arrangement of hydrocarbon chains or aromatic rings). Class *H* isotherms correspond to adsorption of very large molecules (e.g. of polymers) or adsorbable micelles. In the latter case chemisorption may also occur.

NOTE. The adsorption of polymers from solution [40] is an important problem of growing theoretical and practical significance but exceeds the scope of the present book.

4.2.3 METHODS OF MEASUREMENT OF ADSORPTION FROM SOLUTION

We can measure only excess adsorption experimentally using the static or dynamic method. In the static method we measure the difference of concentrations of one of the components, e.g. component 1, in the initial solution and in the solution in equilibrium with the adsorbent.

A simple method of measurement of adsorption from solution consists in immersing a known weight of adsorbent in solutions of different concentrations in closed vessels. When adsorption equilibrium is established (usually after one to several dozen hours), the concentration of the solution over the adsorbent is measured.

If we denote the total number of moles of the solution used for the measurement by n^0 (where $n^0 = n_1^0 + n_2^0$) and the mass of adsorbent used by m, and the mole fractions of component 1 in the initial solution and in adsorption equilibrium by x_1^0 and x_1, then

$$n_1^{\sigma(n)} = \frac{n^0(x_1^0 - x_1)}{m}. \tag{4.54}$$

Now if we know the specific surface area s of the adsorbent, then:

$$\Gamma_1^{(n)} = \frac{n^0(x_1^0 - x_1)}{ms}. \tag{4.55}$$

Similarly we can measure e.g.:

$$n_1^{\sigma\,(v)} = \frac{V(c_1^0 - c_1)}{m} \tag{4.56}$$

and

$$\Gamma_1^{(v)} = \frac{V(c_1^0 - c_1)}{ms} \tag{4.57}$$

where V is the volume of the solution used in the measurement, and c_1^0 and c_1 are volume concentrations (mole per dm^3) of component 1 in the initial and adsorption equilibrium solutions, respectively. If n^0 is expressed in millimoles, V in millilitres, m in grams and s in m^2/g, then $n_1^{\sigma\,(n)}$ and $n_1^{\sigma\,(v)}$ will be given in millimoles per gram, and $\Gamma_1^{(n)}$ and $\Gamma_1^{(v)}$ in millimoles per m^2.

In a more complex method, the solution is pumped through a bed of adsorbent in a closed cycle until adsorption equilibrium is reached. After the concentration of the solution is measured we add to it an accurately measured quantity (e.g. by means of a microburette) of the adsorbed substance. In this way we can calculate the new, higher, initial solution concentration. After the new adsorption equilibrium is established we measure the concentration of the solution, etc.

Figure 4.14 shows a diagram of the apparatus for measuring adsorption by the method described [41]. The column is filled with a known quantity of adsorbent and evacuated at a suitable temperature to 10^{-3} mm Hg. The solution from container *4* is forced through the column containing adsorbent and through one of the flow cells of an interferometer. A solution of the same concentration, obtained using a similar apparatus but without the adsorbent bed, is passed through the second cell of the interferometer.

There are several dynamic (chromatographic) methods of measuring adsorption from solutions, i.e. of determining the excess adsorption isotherms, as in the case of solid/gas interface (see Chapter 3). This can usually be done by frontal analysis, where a solution of concentration c_1 (or x_1) is passed through a column of adsorbent of mass m. If component 1 is more strongly adsorbed than component 2, then initially only the latter is present in the eluate. When the volume of the eluate reaches a certain value V_R referred to as frontal volume (or retention volume), component 1 at concentration c_1 (x_1) appears in it. Often, in view of the diffuse concentration front of component 1 in the adsorbent bed, its concentration in the eluate increases gradually until the value c_1 (or x_1) is reached. In such a case the retention volume V_R is found by interpolation of the outflow curve (Fig. 4.15).

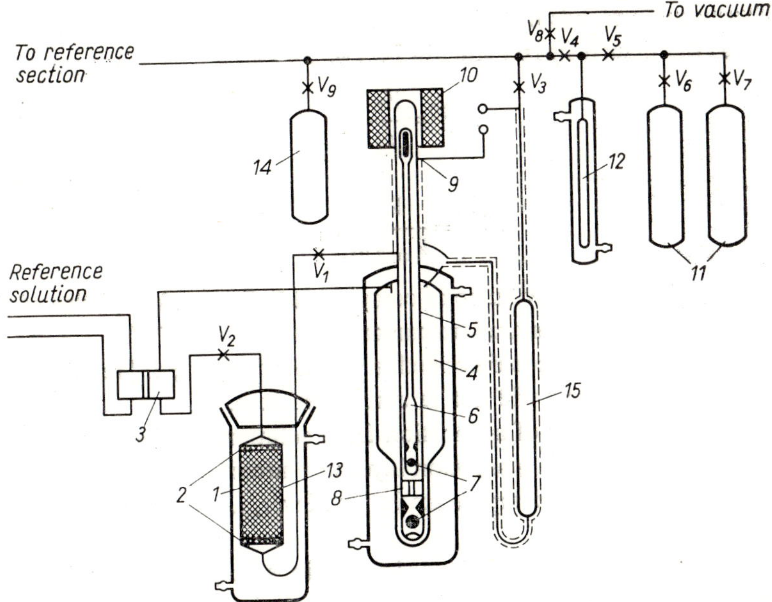

FIG. 4.14 Apparatus for measuring adsorption from a solution by the static method (after Kurbanbekov *et al.* [41]): *1*—column with adsorbent bed, *2*—porous glass, *3*—interferometer with flow cell, *4*—container with the solution, *5*—pump, *6*—piston, *7*—glass valves, *8*—teflon gasket, *9*—iron rod fused into the glass, *10*—coil driving the pump, *11*—ampoules with the solution components, *12*—burette for dosing the solution components into measuring compartment, *13*—sealed opening for filling the column with the adsorbent, *14*—ampoule with solvent, *15*—absorber, *V*—valves.

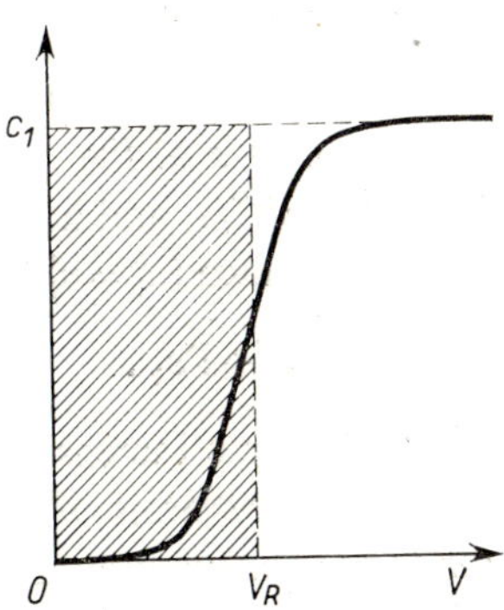

FIG. 4.15 Determination of the retention volume V_R in the dynamic method of measuring adsorption from solutions.

Excess amount adsorption is calculated using the following formula (equal to the shaded surface area in Fig. 4.15):

$$n_1^{\sigma(v)} = \frac{V_R}{m} c_1 \tag{4.58}$$

or

$$n_1^{\sigma(n)} = \frac{V_R}{V_{m,2}\, m} x_1 \tag{4.59}$$

where $V_{m,2}$ is the molar volume of component 2. From a knowledge of the adsorbent specific surface area we can calculate $\Gamma_1^{(v)}$ and $\Gamma_1^{(n)}$.

Other dynamic methods described in the literature [42, 43] require more elaborate apparatus and will not be considered here.

To develop a theory of adsorption from solution and become familiar with the structure of the adsorption layer it is necessary to know the total content of the solution components in that layer, i.e. their individual adsorptions n_1^s and n_2^s.

In accordance with the definitions in equations (4.5) and (4.6), when estimating the individual adsorption of a given solution component, it is necessary to determine the total number of moles n^s of all components in the surface layer or the volume V^s of that layer (if we know the specific surface area s of the adsorbent, it is necessary to measure the thickness l^s of the adsorption layer). Several methods have been developed for determining n^s or V^s, some of which deserve particular attention.

Kiselev and Shcherbakova [44] have estimated the total amount of component 1 in the surface layer of 1 g of the adsorbent from the equation:

$$n_1^{s(v)} = n_1^{\sigma(v)} + V^s c_1 \tag{4.60}$$

where V^s is the volume of the surface layer of 1 g of the adsorbent. From a knowledge of the specific surface area s of the adsorbent we can calculate the thickness of the surface layer $l^s = V^s/s$, and hence the amount of component 1 per unit area of adsorbent surface ($\alpha_1^s = n_1^{s(v)}/s$),

$$\alpha_1^s = \Gamma_1^{(v)} + l^s c_1 . \tag{4.61}$$

The values of $n_1^{s(v)}$ or α_1^s may of course increase with increase of c_1 only up to a certain limit, since the isotherms $n_1^{\sigma(v)} = f(c_1)$ or $\Gamma_1^{(v)} = F(c_1)$ for completely miscible liquids have a maximum and then decrease, often linearly, to zero. When component 1 is strongly adsorbed and component 2 only weakly, the values $n_1^{s(v)}$ or α_1^s tend to a constant value corresponding to the complete filling of the adsorption layer with component 1. The slope of the linearly decreasing part of the isotherms $n_1^{\sigma(v)}$ or $\Gamma_1^{(v)}$ give the values of V^s or l^s:

$$V^s = -\left(\frac{\partial n_1^{\sigma(v)}}{\partial c_1}\right)_{n_{m,1}^s}. \tag{4.62}$$

From the excess adsorption isotherms ($n_1^{\sigma(v)}$ or $\Gamma_1^{(v)}$) we can determine the values of V^s or l^s, and hence we can calculate, using equations (4.60) and (4.61), the isotherms of the total amount ($n_1^{s(v)}$ or α_1^s) of adsorbed component in the surface layer, i.e. the individual adsorption isotherms. Figure 4.16 shows the excess adsorption isotherm of benzene $n_1^{\sigma(v)}$ from *n*-heptane on a wide-porous silica gel as well as the individual adsorption isotherm on the same adsorbent with $l^s = 0.375$ nm [45].

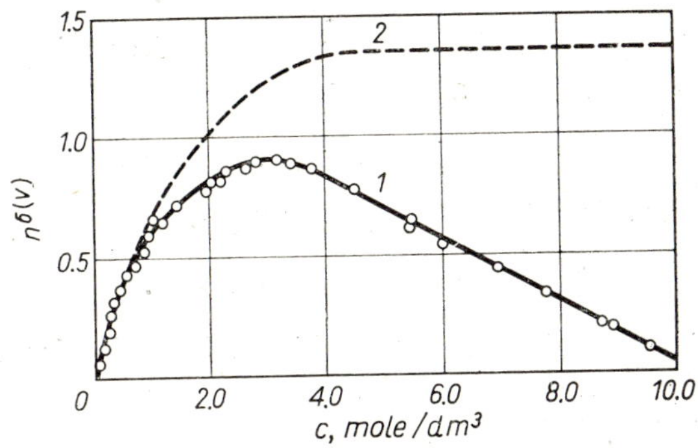

FIG. 4.16 Excess (*1*) and individual (*2*) adsorption isotherms of benzene from *n*-heptane on wide-porous silica gel (after Kiselev and Eltekov [45]).

The relationship between excess adsorption of component 1 from a binary solution and the individual adsorptions n_1^s and n_2^s of the components of that solution is given by equation (4.36), which can be derived as follows.

If n_1, n_2 and n_1^s, n_2^s are the numbers of moles of the components of the bulk and surface solution per 1 g of adsorbent respectively, we can write:

$$n = n_1 + n_2, \quad x_1^s = n_1^s/n^s, \quad n^s = n_1^s + n_2^s,$$

$$n = n^0 - n^s = n^0 - (n_1^s + n_2^s), \quad n^0 = n + n^s.$$

From the mass balance in both phases it follows that:

$$n^0 x_1^0 = n x_1 + n^s x_1^s \tag{4.63}$$

where x_1^0 is the mole fraction of component 1 in the initial solution (prior to adsorption).

In accordance with equation (4.54) with $m=1$ g, we get:

$$n_1^{\sigma(n)} = n^0(x_1^0 - x_1) = n^s(x_1^s - x_1) \tag{4.64}$$

or

$$n_1^{\sigma(n)} = n_1^s - (n_1^s + n_2^s)\,x_1 = n_1^s - n^s x_1\,. \tag{4.65}$$

Equation (4.36) can also be written in the form:

$$n_1^{\sigma(n)} = n_1^s - (n_1^s + n_2^s)\,x_1 = n_1^s\,(1 - x_1) - n_2^s\,x_1$$

or

$$n_1^{\sigma(n)} = n_1^s\,x_2 - n_2^s\,x_1\,. \tag{4.66}$$

The relationship between $n_1^{\sigma(n)}$ and n_1^s, n_2^s, x_1 and x_2 as given by equation (4.66) leads some researchers (Kipling [10]) to refer to excess adsorption isotherms as composite adsorption isotherms.

Schay and Nagy [11, 12] have assumed for the decreasing section of the excess adsorption isotherm that:

$$n_1^{\sigma(n)} = a - bx_1\,. \tag{4.67}$$

By comparing equations (4.65) and (4.67) we get:

$$n_1^s = a \qquad \text{and} \qquad n_1^s + n_2^s = n^s = b\,.$$

Schay and Nagy [46] have found that their assumption is valid for physical adsorption when the dimensions of the molecules of the solution components are similar. The constants a and b in equation (4.67) represent the minimum amounts of components in the adsorption layer if we assume that the composition of that layer is constant within the linear part of the adsorption isotherm (when it is linear within 0.3 mole fraction units) despite changes of composition of the bulk phase. However, this is not completely correct in thermodynamic terms [47].

Analysis of the initial equation indicates that the experimental value of b is smaller than the sum $n_1^s + n_2^s$.

NOTE. According to Schay and Nagy it is assumed that for the linear part of the excess adsorption isotherm we have (equation (4.67)):

$$n_1^{\sigma(n)} = a - bx_1$$

where

$$a = n_1^s \;\text{ and }\; b = -\frac{dn_1^{\sigma(n)}}{dx_1} = n_1^s + n_2^s\,.$$

Considering equation (4.65) we can write:

$$-\frac{dn_1^{\sigma(n)}}{dx_1} = n_1^s + n_2^s + x_1\frac{dn_2^s}{dx_1} - x_2\frac{dn_1^s}{dx_1} = n_1^s + n_2^s + \Delta\,. \tag{4.68}$$

Schay and Nagy assume therefore that $\Delta=0$. Since

$$\Delta=x_1\frac{dn_2^s}{dx_1}-x_2\frac{dn_1^s}{dx_1}=-\frac{dn_1^s}{dx_1}(x_2+x_1\beta^{-1}) \tag{4.69}$$

where $\beta=dn_1^s/dn_2^s$ and $(x_2+x_1\beta^{-1})\neq 0$, it follows from their assumption that:

$$\frac{dn_1^s}{dx_1}=0 \quad \text{and} \quad \frac{dn_2^s}{dx_1}=0$$

and hence also

$$\frac{dx_1^s}{dx_1}=0. \tag{4.70}$$

Equation (4.70) would mean that when the chemical potential of a component of the solution changes, its chemical potential in the adsorption layer remains constant. This indicates that equation (4.70) is inconsistent with the thermodynamical condition of equilibrium. This problem has been considered at length by Rusanov [48] and Bering and Serpinski [49]. The linearity of the decreasing section of the excess adsorption isotherm is connected solely with poor experimental accuracy at higher solution concentrations.

Despite the above limitations, the method of Schay and Nagy gives good results, especially for type II excess adsorption isotherms [46]. Then, for $x_1=1$, we have $n_1^{\sigma(n)}=0$ and hence $b=a$, so that:

$$n_1^{\sigma(n)}=a(1-x_1)=ax_2. \tag{4.71}$$

When the shapes of the molecules of the solution components are almost equal we can draw the relative excess adsorption isotherm:

$$\frac{n_1^{\sigma(n)}}{x_2}=f(x_1) \tag{4.72}$$

If this isotherm is extrapolated to $x_1=1$, in accordance with the suggestion of Schay [50], we get real values of $n_{m,1}^s$ in the adsorption layer (Fig. 4.17).

Another method of calculating the total content of a component in the surface layer has been proposed by Everett [8], and relates to excess adsorption isotherms of types I and II (Fig. 4.5). If we assume that $\beta=1$ in equation (4.43), then:

$$\frac{x_1 x_2}{n_1^{\sigma(n)}}=\frac{1}{n_{m,1}^s}\left(\frac{1}{\alpha-1}+x_1\right). \tag{4.73}$$

For ideal adsorption, $\alpha=K_1=\text{const}$, and then the left-hand side of equation (4.73) is a linear function of x_1. Figure 4.18 illustrates experimental data for adsorption in three systems which support Everett's considerations.

The excess adsorption isotherm of benzene from cyclohexane on silica gel at 298 K is represented in Fig. 4.17 in the normal co-ordinate system $n_1^{\sigma(n)} = f(x_1)$, in the co-ordinate system as given by Schay and Nagy (equation (4.72)), and according to Everett (equation (4.73)). The results of both methods are in this case in good agreement [50].

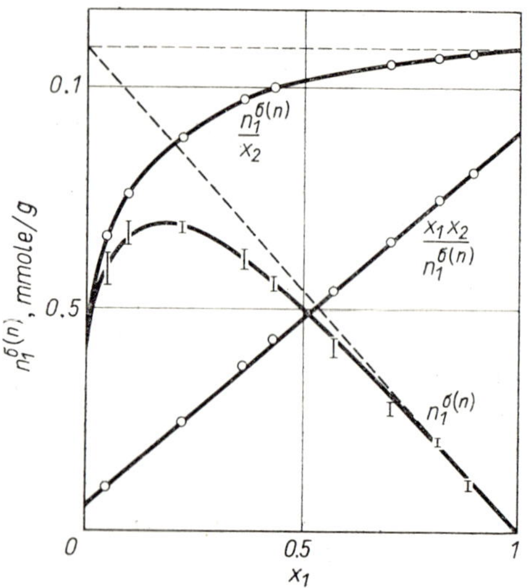

FIG. 4.17 Excess adsorption isotherm $n_1^{\sigma(n)} = f(x_1)$ of benzene from cyclohexane on silica gel [50] in the co-ordinate system as given by Schay and Nagy and by Everett.

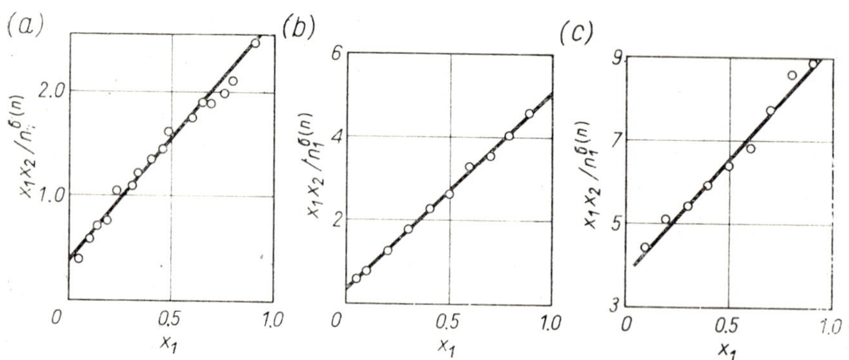

FIG. 4.18 Experimental adsorption data presented according to equation (4.73) (after Everett [8]): (a) benzene–cyclohexane–spheron [51], (b) benzene–cyclohexane–carbon [51], (c) benzene–dichloroethylene–boehmite [52].

If $\beta \neq 1$, then equation (4.43) takes the form:

$$\frac{x_1 x_2}{n_1^{\sigma(n)}} = \frac{1}{n_{m,1}^s}\left(\frac{1}{\beta(\alpha-1)} + \frac{\alpha-\beta^{-1}}{\alpha-1}x_1\right). \tag{4.74}$$

We see that the expression $x_1x_2/n_1^{\sigma(n)}$ will depend almost linearly on x_1 even when $\alpha \neq$const, provided that α is sufficiently large. The first term in the square brackets can then be neglected and x_1 almost equals unity.

Good agreement of experimental data with Everett's equation is often obtained for non-ideal systems. However, there is only formal agreement between experiment and that equation and the results pertaining to the value of $n_{m,1}^s$ do not necessarily represent reality [53].

Laryonov *et al.* [17] have analysed the methods of Kiselev and Shcherbakova and Schay and Nagy and found that each can be used only in the case of strong adsorption of one solution component. It then follows that the excess and individual adsorptions are related by:

$$-\left(\frac{\partial n_1^{\sigma(n)}}{\partial x_1}\right)_{x_1\to 1} = n_{m,1}^s\left(1-\frac{\beta}{K_L^{1/\beta}}\right) \tag{4.75}$$

where by definition:

$$K_L = \frac{x_1^s}{x_1}\left(\frac{x_2}{x_2^s}\right)^\beta . \tag{4.76}$$

From equation (4.75) it follows that these methods give good results when $K_L \gg 1$ (strong adsorption of component 1).

Apart the above graphical methods, we can calculate $n_{m,1}^s$ from the equation:

$$(n_1^{\sigma(n)})_{x_1=x_{1,\max}} = n_{m,1}^s(1-2x_{1,\max}) \tag{4.77}$$

where $x_{1,\max}$ is the mole fraction of component 1 in the equilibrium solution corresponding to the maximum value of $n_1^{\sigma(n)}$ [54].

From a knowledge of $n_{m,1}^s$ (or $n_{m,2}^s$) we can derive individual adsorption isotherms on the basis of, e.g., relation (4.60) ($n_{m,1}^s V_{m,1} = V^s$, where $V_{m,1}$ is the molar volume of component 1).

The calculation of the capacity n^s of the surface layer proposed by the present author and his collaborators [55–58] is based on theoretical considerations of adsorption from binary solutions on non-uniform adsorbent surfaces. Individual adsorption of component 1 both from the gas and liquid phases can be described by an exponential isotherm equation of

Dubinin–Radushkevich type (cf. Chapter 3):

$$x_1^s = x_1 + \frac{n_1^{\sigma(n)}}{n^s} = \exp\left[\sum_{i=1}^{k} B_i \left(RT \ln \frac{x}{x_0}\right)^i\right] \tag{4.78}$$

where $x = x_1/x_2$; $x_0 = x_{1,0}/x_{2,0}$ is a constant related to the maximum difference of the adsorption energies of both components of the solution ($\varepsilon_0 = \varepsilon_{1,0} - \varepsilon_{2,0}$), and the constants B_i are characteristic of a given adsorption system. Equation (4.78) can be generalized to encompass all individual adsorption isotherms of component 1 from binary solutions. Since the mole fraction of component 1 in the surface layer is $x_1^s = n_1^s/n^s$, we can write equation (4.78), after taking logarithms, in the form:

$$\ln n_1^s = \ln n^s + B_1 \ln \frac{x}{x_0} + B_2 (RT)^2 \left(\ln \frac{x}{x_0}\right)^2 + \ldots + B_k (RT)^k \left(\ln \frac{x}{x_0}\right)^k \tag{4.79}$$

from which it follows that $\ln n^s = B_0$ or $n^s = \exp B_0$. The B_i, and hence also B_0, are determined numerically by approximating experimental excess adsorption isotherm data (with an appropriate step with respect to n^s) by the polynomial:

$$y = \sum_{i=0}^{k} B_i' z^i \tag{4.80}$$

where $B_i' = B_i(RT)^i$, $y = \ln n_1^s$ and $z = \ln(x/x_0)$. The approximation is simple and accurate, but attention should be drawn to the criteria of selection of the best polynomial. It allows us to obtain reliable values of the surface layer capacity. Particularly interesting are the cases when $i=1$ (Freundlich's adsorption isotherm) and $i=2$ (classical D–R adsorption isotherm) in equation (4.78).

By means of polynomial (4.80) one can approximate experimental data for all types of excess adsorption isotherms. In Table 4.1 the values of n^s calculated by this method are given for several adsorption systems. For comparison, the values of n^s calculated by Everett's method [8, 9] are also given. The above method of calculating n^s may be particularly useful in cases when the methods of Everett and others cannot be applied.

Methods for calculating the capacity n^s (most frequently $n_{m,1}^s$) of the adsorption layer can serve as a basis for determining the specific surface area of adsorbents from adsorption data for binary solutions, which is

Table 4.1 The capacities n^s of the surface layers ($T=303$ K)

Adsorption system	Calculated capacity of the surface layer	
	according to equation (4.78) mmole/g	after Everett according to equation (4.73) mmole/g
Benzene (1)+cyclohexane (2) Silica gel	2.880	2.777
Benzene (1)+cyclohexane (2) Active carbon	2.151	2.161
Benzene (1)+ethyl acetate (2) Active carbon	1.603	1.616

possible when an adsorption monolayer is formed. Schay and Nagy [46] recommend the use of data from adsorption of benzene from *n*-heptane (or *n*-hexane) as a basis for the determination of the specific surface area of oxide adsorbents (alumina, silica gel, etc.), when type II excess adsorption isotherms are obtained. In such measurements it is assumed that the surface area of the adsorbent occupied by 1 mmole of benzene is 180 m^2. From the capacity $n^s_{m,\,C_6H_6}$ (mmole/g) of the monolayer, we calculate the specific surface area of the adsorbent as follows: $s=180\, n^s_{m,\,C_6H_6}$ (m^2/g).

Schay and Nagy have shown that the specific surface area of carbon adsorbents can be based on adsorption data for lower aliphatic alcohols (methanol, ethanol or *n*-propanol) from benzene. We then obtain type IV excess adsorption isotherms for alcohols. The adsorbent surface area occupied by a monolayer of 1 mmole of alcohol is: 95 m^2 for methanol, 120 m^2 for ethanol and 161 m^2 for *n*-propanol. The adsorbent specific surface area is calculated as for oxide adsorbents.

Table 4.2 Comparison of the specific surface areas (m^2/g) as measured by the BET and Schay and Nagy [46] methods

Adsorbent	Solution	Type of isotherm	s_{BET}	$s_{S\text{-}N}$
Silica gel S 1	benzene+*n*-heptane	II	322	310
Silica gel S 3	benzene+*n*-heptane	II	210	200
Silica gel S 2	benzene+*n*-heptane	II	452	460
Active carbon (Nuxit A II)	ethanol+benzene	IV	770	807
Active carbon	ethanol+benzene	IV	620	587

Schay and Nagy [46] have compared specific surface areas obtained by the BET method with those based on adsorption data from solution (graphical method) for a range of oxide and carbon adsorbents and obtained good agreement (Table 4.2).

According to certain authors [10, 49, 53] adsorption of vapours of solution components provides a good basis for measuring n^s. We assume that the capacities of monolayer adsorption from the gas phase on a given adsorbent correspond to those of the adsorption layer in adsorption from a solution.

4.2.4 Effect of the Adsorbent Structure on Adsorption from Solution

Adsorption of a given substance from solution increases with decreasing size of the adsorbent capillaries provided that the solution components can penetrate the capillaries. Thus narrow-porous adsorbents have greater adsorbing power, as can be seen from the excess adsorption isotherms for benzene from n-heptane solution on wide-porous ($d=10$ nm, $s=340$ m^2/g) and narrow-porous ($d=3.5$ nm, $s=440$ m^2/g) silica gels presented in Fig. 4.19 [45]. The figure also shows individual adsorption isotherms (dashed lines) calculated according to Kiselev and Shcherbakova. For narrow-porous silica gel the excess adsorption isotherm maximum is not only higher but also shifted towards lower equilibrium solution concentrations.

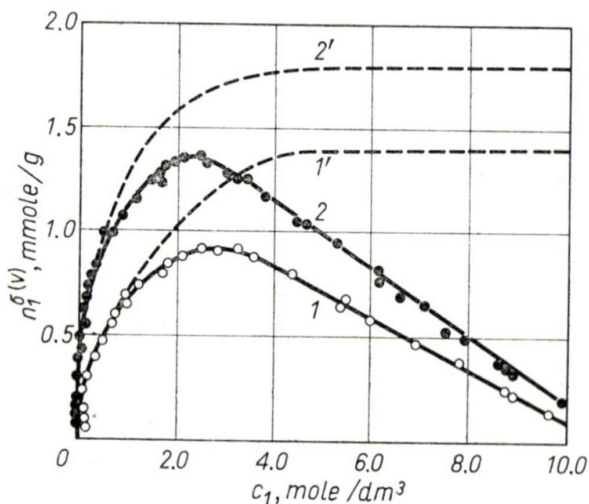

Fig. 4.19 Excess (solid lines) and individual (dashed lines) adsorption isotherms of benzene from n-heptane on wide-porous (*1*) and narrow-porous (*2*) silica gel (after Kiselev and Eltekov [45]).

Comparing the $\Gamma_1^{(v)} = f(c)$ isotherms we shall see that their decreasing linear sections superimpose, confirming that monomolecular adsorption of benzene from *n*-heptane solution takes place on both adsorbents (Fig. 4.20). If the capillary dimensions become smaller than the size of the molecules of one of the solution components, then adsorption of that component is negative even if the nature of the adsorbent surface favours its strong adsorption. This can be observed for adsorption of the components of the benzene–*n*-hexane solution on zeolites.

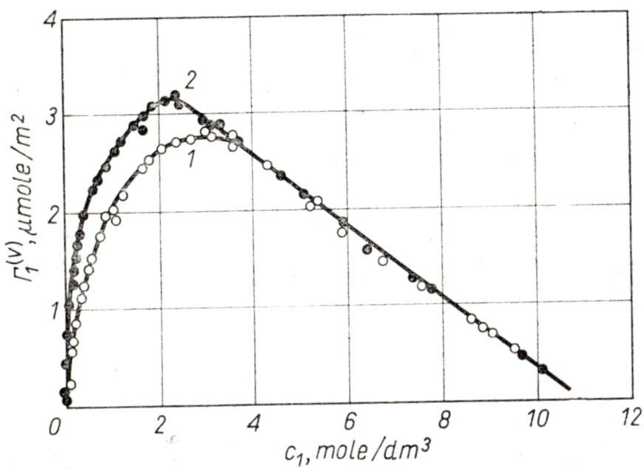

FIG. 4.20 Excess adsorption isotherms of benzene from *n*-heptane on narrow-porous (*2*) and wide-porous (*1*) silica gel (after Kiselev and Eltekov [45]).

Figure 4.21b shows that, from a solution of benzene in *n*-hexane, benzene is very strongly adsorbed on X-type zeolite, although the entrance channels to the inner spaces of this kind of molecular sieve are sufficiently large for both kinds of molecules (benzene and *n*-hexane). Strong adsorption of benzene molecules on a zeolite surface is due to their large affinity with that surface (this will be considered in Chapter 6). However, if the diameters of the zeolite channels are smaller than the dimensions of benzene molecules, then the latter cannot penetrate the channels irrespective of their high affinity towards the adsorbent surface. This is seen in Fig. 21a, where *n*-hexane, not benzene, is adsorbed positively on zeolite 5A from benzene–*n*-hexane solution since the longitudinal molecules of the former can penetrate into the channels of that molecular sieve. In this case negative adsorption of benzene is due solely to geometric factors.

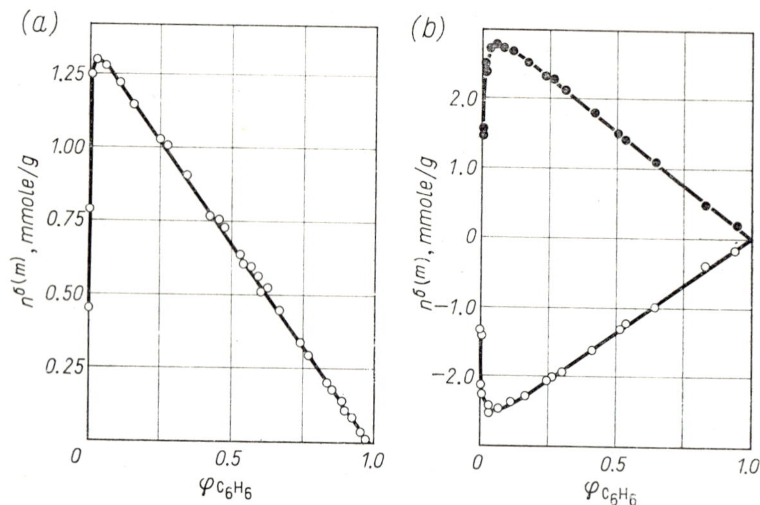

FIG. 4.21 Excess adsorption isotherms of benzene (full circles) and *n*-hexane (open circles) on molecular sieves: (a) 5A, (b) 13X. $\varphi_{C_6H_6}$ – volume fractions of benzene [59, 60].

The above example shows that for the analysis of an adsorption mixture the action of the adsorbent as a molecular sieve is as important as its nature. For this reason, narrow-porous zeolites 4A are used for accurate dehydration of solvents, as only the smallest molecules, especially those of water, can penetrate the adsorbent channels.

4.2.5 CHEMISORPTION

During adsorption of simple organic substances from solution at room temperature, we expect only physical adsorption. Chemisorption is also possible at that temperature, however.

Figure 4.22 represent two excess adsorption isotherms of ethanol from an ethanol–benzene mixture on silica gel at 298 K [11]. Isotherm *2* represents adsorption of ethanol on silica gel pretreated with pure ethanol, so only physical adsorption occurs. Isotherm *1* shows the total adsorption of ethanol, both physical and chemical (when the gel was not modified prior to adsorption). Since these isotherms have linear sections, utilization of the method of Schay and Nagy shows that multilayer adsorption occurs (the first layer being chemisorbed) and that the amount of ethanol physically adsorbed is somewhat greater than half the total amount adsorbed.

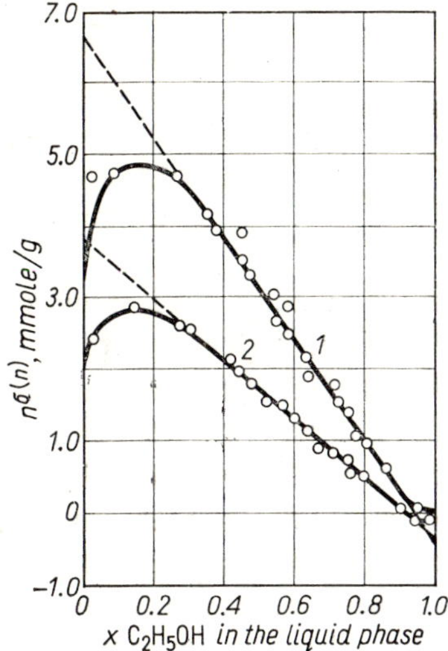

FIG. 4.22 Excess adsorption isotherms of ethanol from benzene on non-modified gel *1* and gel preliminarily treated with pure ethanol *2* (after Schay *et al.* [11]).

When chemisorption occurs, equation (4.66), representing the surface excess amount of component 1 as a function of bulk solution composition and number of moles of both components in the surface solution, should be modified as follows:

$$n_1^{\sigma(n)} = [(n_1^s)_{chem} + (n_1^s)_{phys}]\, x_2 - n_2^s x_1 \tag{4.81}$$

where $(n_1^s)_{chem}$ and $(n_1^s)_{phys}$ are the numbers of moles of component 1 adsorbed chemically and physically (per unit mass of the adsorbent). The term $(n_1^s)_{chem}$ is independent of bulk solution composition, and $(n_1^s)_{phys}$ depends on x_1 as defined by equations (4.40) and (4.42).

4.2.6 Orientation of Adsorbed Molecules

The orientation of adsorbed molecules (i.e. molecules comprising the surface solution) depends both on the nature of the interaction between the adsorbent surface and these molecules and the nature of the interactions of molecules comprising the bulk solution. The nature of molecular interactions in the adsorption process will be discussed in detail in Chapter 6. We shall therefore only comment here on the orientation of molecules adsorbed in solution adsorption.

In chemisorption the orientation of molecules in the surface layer presents no problem, especially for monofunctional adsorbates, the functional group being that part of the molecule which adheres to the adsorbent surface. In that case, for instance, the molecules of fatty acids with long chains are bonded to the adsorbent by their carboxylic group. The remaining part of the molecule is usually oriented vertically with respect to the surface.

A similar orientation of molecules on adsorbent surface is often assumed for physical adsorption, while, in reality, a different orientation of the molecules frequently occurs. For instance, when stearic acid is adsorbed on active carbon from cyclohexane the adsorbed molecules are arranged parallel to the adsorbent surface. Kipling and Wright [61, 62] have found that the arrangement of molecules of other fatty acids on a number of non-polar adsorbents is similar, this arrangement being due to the formation of dimers in such solvents as benzene, and the non-polar adsorbent does not have a sufficient adsorbing strength to split these dimers. On polar adsorbents such as alumina and titania, fatty acid molecules have been found to be arranged vertically with respect to the adsorbent surface [63, 64]. Paraffin hydrocarbons always assume a parallel orientation with respect to any adsorbent surface [10].

The studies of Daniel [65] reveal the possibility of a change of orientation of the adsorbate molecules on adsorbent surface with variation of the

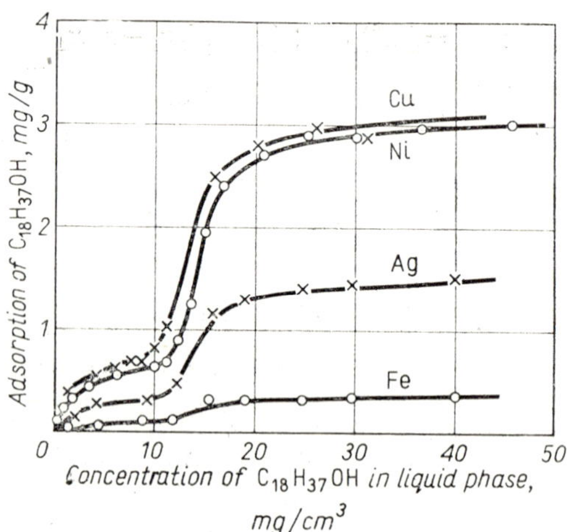

FIG. 4.23 Adsorption isotherms of octadecanol from benzene on powdered metals (after Daniel [65]).

equilibrium solution concentration. The adsorption isotherms of octadecanol from benzene on powdered metals show strong bends (Fig. 4.23). The shape of these isotherms indicates that at low solution concentrations (small coverage of adsorbent surface with alcohol molecules) the adsorbed alcohol molecules are arranged parallel to, and at higher concentrations (greater coverage of adsorbent surface) vertically to, the adsorbent surface. It has been found, however, that on other adsorbents (silica gel) adsorbed octanol molecules are also oriented parallel to the surface at higher equilibrium solution concentrations.

Hence decisive factors for this or that orientation of adsorbate molecules in the course of adsorption from solution are very complex. In the absence of detailed knowledge of the nature of intermolecular interactions between the adsorbent surface and the molecules of the solution components it is not possible to lay down precise quantitative rules to explain the orientation of molecules in the surface solution.

4.2.7 Adsorption from Multicomponent Solutions

Adsorption from multicomponent solutions is a very complex process and presents many difficulties. It can be aproached in two ways: (i) as the adsorption of a given substance from a multicomponent solvent, or (ii) as the adsorption of many substances from a given solvent.

We shall consider only the first approach here as it is more widely represented in the literature.

Gryazev [66] studied experimentally adsorption from ternary systems over a fairly wide range of concentrations. The spatial excess adsorption isotherms of acetic acid from a ternary acetic acid–lauric acid–cetane solution on Volga rocks obtained in these studies are illustrated in Fig. 4.24.

The present author [67–69] has considered the thermodynamics of adsorption from multicomponent solutions. He found that for ideal and regular systems when the ratio of mole (or volume) fractions of the components of a binary solution $k+i$ is, under adsorption equilibrium conditions, equal to the ratio of mole (volume) fractions of the same components in a multicomponent solution Σ_i, i.e. when:

$$\left(\frac{x_k}{x_i}\right)_{i,k}=\left(\frac{x_k}{x_i}\right)_{\Sigma_i} \tag{4.82}$$

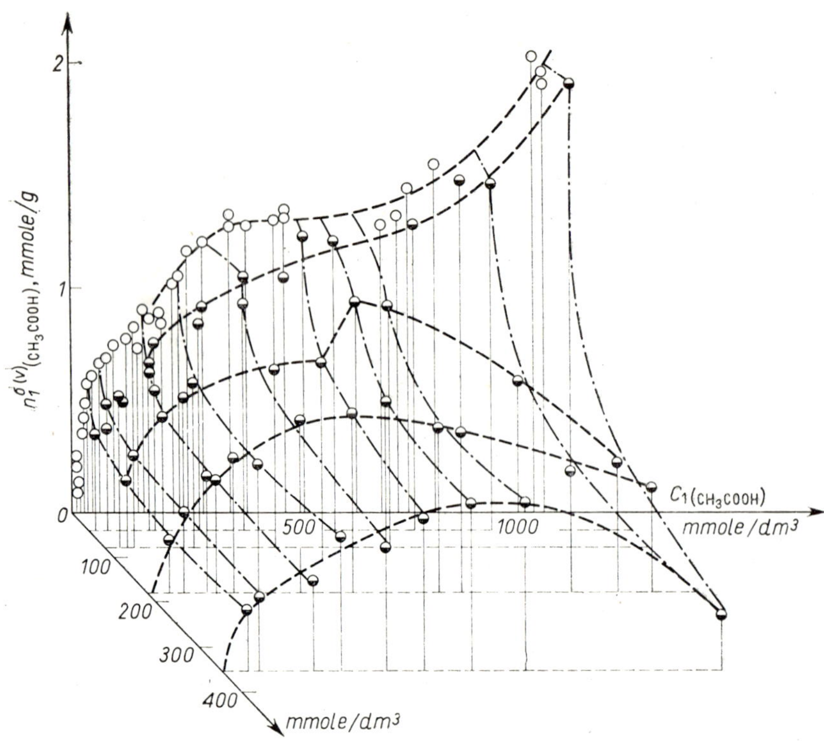

FIG. 4.24 Spatial excess adsorption isotherms of acetic acid from ternary acetic acid–lauric acid cetane solution on active Volga rocks: adsorption of acetic acid from cetane (open circles), adsorption of acetic acid from the ternary system (semifull circles) (after Gryazev [66]).

then the ratios of the mole (volume) fractions of these components in the surface layer are also equal, i.e.

$$\left(\frac{x_k^s}{x_i^s}\right)_{i,k}=\left(\frac{x_k^s}{x_i^s}\right)_{\Sigma_i}. \tag{4.83}$$

It follows that if condition (4.82) is fulfilled, we have:

$$(x_k^s)_{\Sigma_i}=\Big[\sum_{i\neq k}^{n-1}(x_k^s)_{i,k}^{-1}-(n-2)\Big]^{-1} \tag{4.84}$$

where n is the number of components of the multicomponent solution.

Equation (4.84) allows us to calculate the mole fraction of the given substance (adsorbed from an ideal or regular multicomponent solution) in the surface layer on the basis of its adsorption isotherms from suitable binary solutions. Figure 4.25 shows isotherms, measured and calculated according to the transformed equation (4.84), of excess adsorption of aniline on silica gel from the ternary solution: aniline–carbon tetrachloride–chloroform for three different volume ratios.

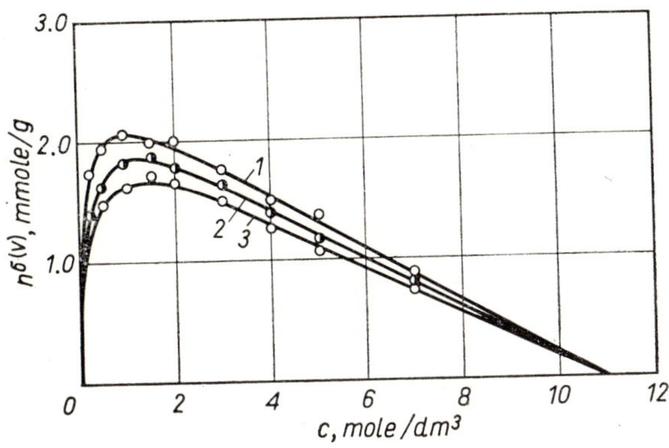

FIG. 4.25 Excess adsorption isotherms of aniline from the ternary aniline–carbon tetrachloride–chloroform solution for three different $CCl_4/CHCl_3$ volume ratios: *1*–3 : 1, *2*–1 : 1, *3*–1 : 3. The solid lines represent isotherms calculated from equation (4.84), and the circles – the experimental points.

Equation (4.84) may also serve as a good approximation for more complex systems in which no strong specific molecular interactions between the molecules of the system components occur.

Kiselev *et al.* [70, 71] studied adsorption from ternary solutions: dioxane–*n*-hexane–benzene on silica gel. The studies were conducted by the static method making use of gas chromatography for analysis of the composition of the equilibrium adsorption solution. The excess adsorption isotherm of the given component from the ternary solution represents the surface area in the co-ordinate system: adsorption and composition of equilibrium solution. This surface area can be represented as the projection of the equal adsorption line on to the Gibbs triangle. The adsorption of dioxane and benzene in that system is presented in Figs. 4.26a and 4.26b. The equilibrium solution composition is given in weight fractions and the excess adsorption is $\Gamma_i^{(m)}$. Similar investigations have been carried out [72] for the benzene–heptane–cyclohexane system with silica gel as adsorbent.

Minka and Myers considered theoretically [73] adsorption from a ternary solution, relating these considerations to the data for adsorption from the relevant binary solutions (distribution function α). Experimental

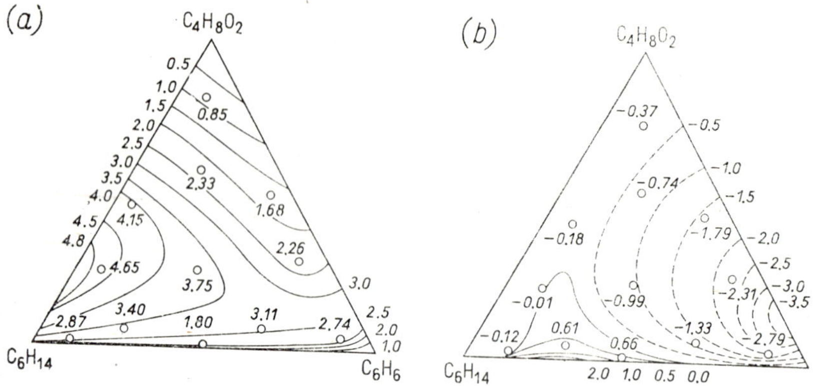

FIG. 4.26 Adsorption of the dioxane–*n*-hexane–benzene ternary solution on silica gel (after Kiselev *et al.* [70, 71]): the projections are given of spatial excess adsorption isotherms on to the Gibbs triangle for dioxane (a) and benzene (b). The solid line presents positive and the dashed one negative adsorption.

studies related to adsorption on active carbon from the benzene–ethyl acetate–cyclohexane ternary solution. The calculated excess adsorption isotherms for component A of an A+B+C solution have been presented at equal mole fractions of the remaining components in the equilibrium solution, $x_B - x_C$. The results were compared with experimental data and with the relevant excess adsorption isotherms of component A from binary A+B and A+C solutions.

The complexity of adsorption from multicomponent solutions together with the great difficulties of studying it have produced relatively few scientific works in this field.

4.3 Adsorption from Solution on Heterogeneous Adsorbent Surfaces

4.3.1 Introduction

The majority of studies of adsorption from solution assume for theoretical simplicity that the adsorbent surface is energetically homogeneous.

The first quantitative description of adsorption from solutions on heterogeneous surfaces was made by Schuchowitzky [74] who, using the potential theory of adsorption from a gas mixture and introducing the Langmuir adsorption model, obtained the excess adsorption isotherm which was later derived under different assumptions (see equation (4.43)). That equation can be written in the form (when $\beta=1$):

$$n_1^{\sigma(n)}=\frac{n^s x_1 x_2(\alpha-1)}{\alpha x_1+x_2} \tag{4.85}$$

where α is the distribution function:

$$\alpha=C\exp\left(\frac{\varepsilon}{RT}\right). \tag{4.86}$$

The quantity $\varepsilon=\varepsilon_1-\varepsilon_2$ is the difference of adsorption potentials of the "1+2" solution components.

NOTE. It was previously shown that:

$$\alpha=C\exp\left[-\frac{1}{RT}(\Delta_a\mu_1^\ominus-\Delta_a\mu_2^\ominus)\right]$$

where: $\Delta_a\mu_1^\ominus=\mu_1^{\ominus,s}-\mu_1^\ominus$ and $\Delta_a\mu_2^\ominus=\mu_2^{\ominus,s}-\mu_2^\ominus$ are the standard chemical potentials of adsorption (differential molar Gibbs free adsorption energies) of the binary solution components. Therefore it follows that:

$$\varepsilon_2-\varepsilon_1=\Delta_a\mu_1^\ominus-\Delta_a\mu_2^\ominus.$$

The above difference is of course constant, characteristic of the given system, in the case of a homogeneous adsorbent surface.

Schuchowitzky has assumed that there are two types of active centres on the adsorbent surface: those at which only component 1 is adsorbed, and those at which only component 2 is adsorbed. The excess adsorption isotherm can then be written in the form of a two-term equation:

$$n_1^{\sigma(n)}=\frac{R_1(\alpha_1-1)x_1x_2}{\alpha_1x_1+x_2}+\frac{R_2(\alpha_2-1)x_1x_2}{\alpha_2x_2+x_1} \tag{4.87}$$

where R_1 and R_2 are numbers of adsorption centres of type 1 and 2, respectively, per unit mass of adsorbent. By fitting equation (4.87) to experimental data we can determine the parameters α_1 and α_2 as well as R_1 and R_2. Schuchowitzky has also suggested that the sign of the excess adsorption isotherm is related to the adsorbent surface heterogeneity.

Šiškova and Erdös [18–20] have derived equation (4.85) from the thermodynamic equilibrium condition. Like Schuchowitzky, they assumed that two types of adsorption centres occur on the adsorbent surface and calculated the fractions of the surface area related to the adsorption potentials ε_1 and ε_2.

Everett [8, 9] also underlined the significant role of surface heterogeneity of adsorbents in adsorption from solutions. Coltharp and Hackerman [75] have shown experimentally in their studies of adsorption from solutions on active carbon that the adsorbent surface heterogeneity may be the cause of the change in sign of excess adsorption of one of the components of a liquid mixture.

Kagiya *et al.* [76, 77] have studied adsorption on silica gels and confirmed the existence of three types of active sites on the surfaces of these adsorbents, determining their quantitative contributions to the adsorption process.

Studies of adsorbent surface heterogeneity effect on adsorption from solution are only qualitative or semiquantitative. No fully quantitative theory exists and the effect on experimental adsorption data cannot be predicted.

4.3.2 Integral Adsorption Isotherm from Solution

New studies of the mechanism of adsorption from solution on heterogeneous surfaces of solids have been initiated by the present author and his collaborators ([55–58] and [78–82]). The integral isotherm was applied to adsorption from solution and a patch model of adsorbent surface heterogeneity was assumed (cf. Chapter 3).

The description of the adsorption process from solution will be considered for the case of a two-component liquid mixture with approximately equimolar volumes of both components. The magnitude and selectivity of adsorption as well as the influence of the energetical structure of the adsorbent surface on adsorption are determined by the adsorption energy difference $\varepsilon=\varepsilon_1-\varepsilon_2$. If we denote the left-hand side of equation (4.85) by $n_{1,1}^{\sigma(n)}(x_1, \varepsilon)$, it can be regarded as the excess adsorption isotherm of component 1 on the given homogeneous patch of adsorbent surface. We then write n_1^s instead of n^s to denote the capacity of the adsorption monolayer on the given patch. Using the integral adsorption isotherm we can write:

$$n_1^{\sigma(n)}(x_1)=\int_{\Delta}\left[\frac{x_1 x_2(\alpha-1)}{\alpha x_1+x_2}\right]\chi(\varepsilon)\,\mathrm{d}\varepsilon \tag{4.88}$$

where $\Delta=(-\infty, +\infty)$ denotes the range of ε. In equation (4.88) the energy distribution function $\chi(\varepsilon)$ fulfils the following normalization condition:

$$\int_{\Delta} \chi(\varepsilon)\, d\varepsilon = n^s. \tag{4.89}$$

NOTE. The quantities n_1^s are of course allowed for in equation (4.88) by the function $\chi(\varepsilon)\, d\varepsilon$.

The individual adsorption isotherm of component 1 (equation (4.17)) can be written in the form:

$$x_1^s = \left(1 + \frac{1}{\alpha}\frac{x_2}{x_1}\right)^{-1}. \tag{4.90}$$

Considering equation (4.86) and writing $C' = 1/C$ and $x = x_1/x_2$ we get:

$$x_1^s = \left[1 + \frac{C'}{x}\exp\left(\frac{-\varepsilon}{RT}\right)\right]^{-1}. \tag{4.91}$$

For monomolecular adsorption, the coverage $\theta_1(x)$ corresponds to x_1^s. Denoting the left-hand side of equation (4.91) by $\theta_{1,1}(x, \varepsilon)$ we may treat it as the local individual adsorption isotherm of component 1 on the given homogeneous patch of adsorbent surface. Then the individual integral adsorption isotherm has the form:

$$\theta_1(x) = \int_{\Delta} \left[1 + \frac{C'}{x}\exp\left(\frac{-\varepsilon}{RT}\right)\right]^{-1} \chi(\varepsilon)\, d\varepsilon. \tag{4.92}$$

This equation formally resembles the integral adsorption isotherm for adsorption from the gas phase (cf. Chapter 3). Note that the function $\chi(\varepsilon)$ is normalized to unity.

Equation (4.92) gives a range of possibilities for studying the effects of adsorbent surface inhomogeneity on adsorption from solution. It allows us, for instance, to obtain an analytical form of the adsorption isotherm $\theta_1(x)$ for previously assumed analytical forms of $\chi(\varepsilon)$ [78]. It also permits the study of the variation of the distribution function $\chi(\varepsilon)$ with assumed form of adsorption isotherm [80, 81]. Finally it is possible to obtain a numerical solution of the distribution function $\chi(\varepsilon)$ by series expansion of the functions $\theta_1(x)$, $\theta_{1,1}(x, \varepsilon)$ and $\chi(\varepsilon)$ in terms of a complete orthogonal set of generating functions.

The present author and his collaborators [81] have shown that the most general form of individual adsorption isotherm of component 1 can be written as:

$$\theta_1(x) = \exp\left[\sum_{i=1}^{k} B_i \left(RT \ln \frac{x}{x_0}\right)^i\right] \tag{4.93}$$

for $x \ll x_0$, where x_0 is a parameter related to the maximum difference of adsorption energies of components 1 and 2 ($\varepsilon_0 = \varepsilon_{1,0} - \varepsilon_{2,0}$), and B_i are constants analogous to those of the Dubinin–Radushkevich equation (3.45). Equation (4.93), formally similar to the equation describing adsorption from the gaseous phase, can also be considered as adsorption isotherm of D–R type with

$$\varepsilon_0 = RT \ln \frac{C'}{x_0}. \tag{4.94}$$

To obtain the function corresponding to equation (4.93), equation (4.92) should be reduced to the Stieltjes transform [83] by suitable substitutions. Using the known formulae foı inverting the Stieltjes transform, we obtain the following analytical adsorption energy distribution function:

$$\chi(\varepsilon) = \left[\sum_{i=1}^{k} iB_i(\varepsilon_0 - \varepsilon)^{i-1}\right] \exp\left[\sum_{i=1}^{k} B_i(\varepsilon_0 - \varepsilon)^i\right]. \tag{4.95}$$

Equation (4.95) can also be derived using a variational procedure in which the local adsorption isotherm $\theta_{1,1}(x, \varepsilon)$ is approximated by a suitable step function, simultaneously taking account of equation (4.85). From the experimental excess adsorption isotherm data we can calculate the B_i as well as the numerical value n^s of the surface layer capacity in the manner described in Section 4.2.3. Knowledge of these quantities allows us in turn to determine the energy distribution function $\chi(\varepsilon)$ and the theoretical individual adsorption isotherms of the given component of the solution. In Fig. 4.27 we illustrate the adsorption energy distribution function $\chi(\varepsilon)$, and in Fig. 4.28 we show the theoretical excess adsorption isotherm for the benzene–cyclohexane–silica gel system studied by Sircar and Myers [84] assuming an homogeneous adsorbent surface. Note the agreement of the theoretical excess adsorption isotherm of benzene with experiment under the assumption of heterogeneity of the silica gel surface. Agreement with experiment is better on the initial segment of the isotherm, i.e. in the region where the influence of the adsorbent surface heterogeneity on adsorption is greatest. The plot of the $\chi(\varepsilon)$ function points to the existence of two main types of active sites on the silica gel surface (peaks). A considerable part of the $\chi(\varepsilon)$ curve lies in the range of positive values of ε, connected with the stronger adsorption of benzene on silica gel. A part of the energy distribution function plot lies in the range of negative ε values which suggests the existence of active sites on the adsorbent surface where the adsorption of cyclohexane is stronger.

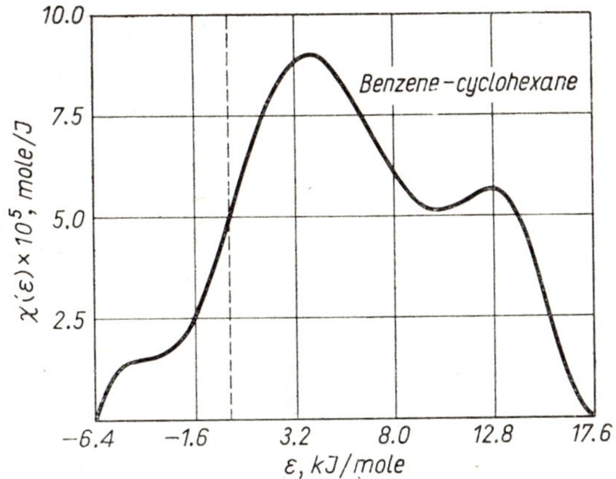

FIG. 4.27 Adsorption energy distribution function $\chi(\varepsilon)$ plot for the silica gel–benzene–cyclohexane system (after Ościk *et al.* [56]).

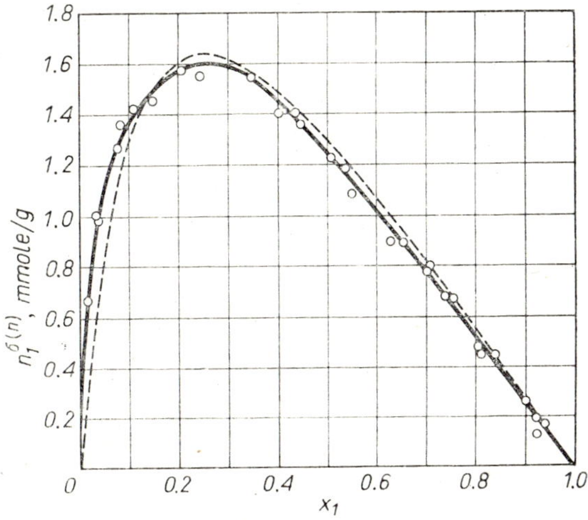

FIG. 4.28 Excess adsorption isotherms of benzene from cyclohexane on silica gel. The solid line presents the isotherm calculated from equation (4.88) (after Ościk *at al.* [56]). The dashed line presents the isotherm calculated by Sircar and Myers [84] assuming energy homogeneity of the adsorbent surface. Open circles present experimental points.

NOTE. To avoid ambiguity, emphasis should be placed on the difference between adsorption energy, $\varepsilon = \varepsilon_1 - \varepsilon_2$, and the adsorbent surface energy, $E = q - q_0$, referred to the ground state q_0. It has been shown [81] that the adsorbent surface energy is related to the temperature of adsorbent treatment and characterizes the adsorbent itself and the heterogeneity of its surface. In contrast, the energy E is independent of the nature of binary liquid mixture in contact with the solid. There exists a relationship between the two kinds of energy, ε and E, which can sometimes be represented analytically.

Modern methods of investigation of adsorbent surface heterogeneity form only a fragment of more extensive studies, which concern, for instance, analogies between adsorption on heterogeneous adsorbent surfaces from the gaseous and liquid phases [56] as well as the mechanism of multilayer adsorption on such surfaces from solutions [58]. It is possible to use a model different from the patch model and the virial formalism can be applied to the description of adsorption from solution on to heterogeneous solid surfaces [85, 86].

4.4 Ion Exchange Adsorption

4.4.1 Introduction and Fundamentals

More than a century ago it was found that some natural minerals such as aluminosilicates from the zeolite group as well as various clays of montmorillonite or bentonite type have the capacity of exchanging ions with solutions. The ion exchange process occurs at the solid/electrolyte solution interface and is therefore referred to as ion-exchange adsorption.

Solids insoluble in water and many other solvents which are able to exchange ions with the solution are known as ion exchangers; those which exchange anions (cations) are called anion (cation) exchangers.

The number of ions which can be exchanged per unit mass or unit volume of the ion exchanger is called the exchange capacity, expressed in gram-equivalents per 1 kg or in milli-gram-equivalents per 1 g of ion exchanger.

The total exchange capacity is the number of milli-gram-equivalents of ions which under most advantageous conditions can be exchanged for other ions by unit mass or unit volume of the exchanger.

The effective exchange capacity is the practical value, always smaller than the total exchange capacity and characteristic of the given exchange process since it depends on the solution concentration, nature of ions involved, temperature, time of contact of the solution with the exchanger, etc.

Ion exchange may also occur at the aqueous solution/organic liquid interface when the organic liquid has a tendency to form bonds with the ions present in the aqueous solution. In the organic phase one can also dissolve substances capable of yielding complex compounds with inorganic ions. Such organic phases or liquid organic substances are called liquid ion exchangers.

4.4.2 Types of Ion Exchangers

The simplest means of classifying ion exchangers is, following historical development, according to their derivation, viz. into (i) natural, (ii) semisynthetic and (iii) synthetic ones.

Natural ion exchangers. The ion exchangers which first found practical application (in the demineralization of water) were cation exchangers of mineral origin (zeolites), which are primarily calcium and sodium aluminosilicates behaving similarly to free silicic acids. Zeolites have the general form

$$(Me^{2+}, Me_2^{+})O;\ Al_2O_3 \cdot nSiO_2 \cdot mH_2O$$

This group includes the minerals analcime, chabazite, natrolite, scolecite, etc. which have a three-dimensional crystalline lattice with SiO_4 and AlO_4 tetrahedra as basic elements of the skeleton. These tetrahedra are condensed to form tetra- or hexamembered rings. An example of zeolite structure is shown in Fig. 3.39 (p. 93). The aluminium ions Al^{3+} carrying a triple positive charge are surrounded in the tetrahedra by oxygen ions with double negative charge or OH^- groups with a single negative charge. Thus the aluminosilicate skeleton has an excess of negative charges which are balanced by alkali metal Me^+ or alkaline earth Me^{2+} ions not included in the skeleton but loosely bound in the channels of the crystal lattice. These ions, being anti-ions, can be displaced and therefore substituted by other ions such as those from the surrounding solution. Owing to the arrangement of open channels existing in zeolites the latter can also operate as molecular sieves as has been discussed in Chapter 3.

Among other natural ion exchangers are montmorillonite, glauconite, and fertile soils which often act as amphoteric ion exchangers.

Semisynthetic ion exchangers. These are ion exchangers of natural origin which have been subjected to suitable chemical treatment. This group is characterized by the so-called sulphonated coals obtained by sulphonation of natural coal with concentrated or fuming sulphuric acid. It has been found [87] that cation exchangers of very high exchange capacity can be obtained from certain kinds of coal. Sulphonated coals are marketed under the trade names of Zeo-Karb-H, Permutit H, Wofatit Z, Escarbo H, etc.

Synthetic ion exchangers. The first synthetic cation exchangers were obtained by inorganic synthesis; they were of aluminosilicate type but of diverse chemical composition depending on the nature of the raw materials used and the production process applied. Such aluminosilicates have the general composition:

$$Al_2O_3 \cdot (SiO_2)_x \cdot (Na_2O)_y \cdot (H_2O)_z$$

Among the ion exchangers currently available, the most valuable are those obtained by organic synthesis. These are synthetic resins, insoluble in water and some organic solvents such as alcohols, ethers, hydrocarbons, etc. Their high tensile strength and chemical resistance as well as high exchange capacity make them suitable for a diversity of applications.

Resins with ion exchange ability differ from plastics, since they have chemically active (functional) groups built into the polymer skeleton. These groups have ionic bonds while the bonds in the skeleton of the macromolecule resin are atomic. The composition of an ion exchange resin can be represented as:

$R—A^-Me^+$	$R—B^+X^-$
cation exchanger	anion exchanger

where R is the polymer skeleton, A^- is the anionic group (acidic, e.g. $—COO^-$) bonded to the polymer by an atomic bond, Me^+ is a cation (H^+ or metal ion) bonded to the group A by an ionic bond and capable of dissociation, B^+ is the cationic group (basic, e.g. $=NH_2^+$) bonded to the polymer by an atomic bond, X^- is an anion (e.g. OH^-) bonded by an ionic bond to the group B and capable of dissociation.

For simplicity only one functional group, $—A^-Me^+$ or $-B^+X^-$, has been shown in the above formulae, while in practice the resin skeleton contains many such groups. The ion exchanger is therefore a polyelectrolyte in which one species of ion constitutes an insoluble skeleton in the form of a porous gel, and the second species of ion with opposite sign is mobile and capable of diffusion (over a certain distance) and therefore of being exchanged for other ions.

Cation exchangers usually have the following specific groups:

sulphonic	$(—SO_3H)^-H^+$
carboxylic	$(—COO)^-H^+$
phenolic	$(—O)^-H^+$
tiophenolic	$(—S)^-H^+$

The most common functional groups in anion exchangers are:

prim. amine	$(-NH_3)^+OH^-$
sec. amine	$(=NH_2)^+OH^-$
tert. amine	$(\equiv NH)^+OH^-$
quater. ammonium	$(\equiv N)^+OH^-$

Ion exchangers of organic origin have a highly developed surface. The solvent and ions can diffuse freely from the surrounding solution into the bulk of the exchanger lattice and react with ions of the functional groups.

4.4.3 The Ion Exchange Process

Ion exchange occurring on a cation exchanger RMe_2 can be represented by the formula:

$$RMe_2 + Me_1X \rightleftarrows RMe_1 + Me_2X \tag{4.96}$$

where R is ion exchange resin with the functional group, Me_2 is the mobile cation liable to exchange, and Me_1X is the electrolyte in solution. The analogous exchange reaction on an anion exchanger proceeds according to the formula:

$$RHX_2 + MeX_1 \rightleftarrows RHX_1 + MeX_2 \tag{4.97}$$

where RH is the anion exchanger with the functional group, and X_2 is the mobile anion liable to exchange.

Ion exchange is therefore a reversible reaction which, depending upon the concentration, properties of the ions and nature of the exchanger, can proceed in one direction or the other.

When ion exchange proceeds in a static system, equilibrium is established between the ions in the solid phase and in solution. This equilibrium conforms with the law of mass action, the Langmuir or Freundlich adsorption isotherms or similar relationships.

Reactions (4.96) or (4.97) can be written as follows:

$$R-M_2 + M_1 \rightleftarrows R-M_1 + M_2 \tag{4.98}$$

In accordance with the law of mass action we can write:

$$K = \frac{a_{M_2 i}\, a_{M_1}}{a_{M_1 i}\, a_{M_2}} \tag{4.99}$$

where $a_{M_1 i}$ and $a_{M_2 i}$ are the activities of ions M_1 and M_2 in the ion exchanger, a_{M_1} and a_{M_2} are the activities of the same ions in solution.

The activity of ions in the exchanger is often given in terms of the number of ions (number of gram-equivalents) per unit mass of exchanger. This allows us to calculate the equivalent fractions $X_{M_1 i}$ and $X_{M_2 i}$ in the inner ion exchanger phase (where $X_{M_1 i}+X_{M_2 i}=1$). Consequently, equation (4.99) becomes

$$\frac{X_{M_2 i}}{X_{M_1 i}} \frac{C_{M_1}}{C_{M_2}} \frac{f_{M_1}}{f_{M_2}} = K \frac{f_{M_1 i}}{f_{M_2 i}} \tag{4.100}$$

where C_{M_1}, C_{M_2} and f_{M_1}, f_{M_2} are the concentrations and activity coefficients of ions M_1 and M_2 respectively in the solution.

All quantities on the left-hand side of equation (4.100) can be determined experimentally for practical purposes, however, it is useful to introduce the apparent equilibrium constant K_a defined by:

$$\frac{X_{M_2 i}}{X_{M_1 i}} \frac{C_{M_1}}{C_{M_2}} \frac{f_{M_1}}{f_{M_2}} = K_a \left(= K \frac{f_{M_1 i}}{f_{M_2 i}} \right). \tag{4.101}$$

If we now neglect the activity coefficient in solution, we get:

$$\frac{X_{M_2 i}}{X_{M_1 i}} \frac{C_{M_1}}{C_{M_2}} = K_c . \tag{4.102}$$

For low ion concentrations in solution $K_a = K_c$.

Experimental data obtained by numerous authors show that K_a is not constant when the ratio $X_{M_2 i}/X_{M_1 i}$ varies. This shows that the inner ion exchanger phase cannot be considered to be an ideal solution, and K_a can be assumed to be constant only over a certain range.

The ion exchange process proceeding in a dynamic system has a different character. Here, the ion being removed leaves the reaction medium, and the exchange process can lead to a complete removal of the ions initially bound to the exchanger.

The dynamic exchange process usually proceeds in an ion exchange bed through which is passed a solution containing the ion to be exchanged. If a sodium chloride solution is passed through a cation exchange bed whose functional groups are substituted by hydrogen ions, then we can distinguish three lateral zones in the bed (Fig. 4.29): the upper one, susbstituted with Na^+ ions, called the post-exchange zone (A); the middle zone, where exchange is in progress (both Na^+ and H^+ ions are bonded to the exchanger functional groups), called the exchange zone (B); the third, lower zone, not yet reached by the NaCl solution, referred to as pre-exchange zone (C). In the exchange zone (B) of the bed both kinds of ions (sodium and hydrogen) are present both in the exchanger and solution. However, the concentration of particular ions varies at different sites of that zone,

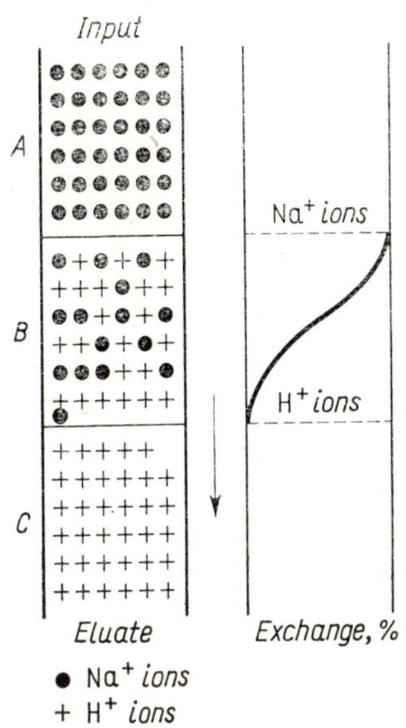

FIG. 4.29 Diagram of the ion exchange in a cation exchanger column.

and the solution composition is a function of the distance from the top of the exchange bed; the ratio c/c_0 (or exchange percentage) as a function of that distance in given time (the so-called isochrone of exchange) is also represented in Fig. 4.29. The eluate will be free from sodium ions and contain an equivalent number of hydrogen ions until the quantity of solution passed through the bed does not produce any shifting of the exchange zone to the very bottom of the ion exchange bed. After a definite amount of solution has passed through the bed, the break-through point is reached, when sodium ions start appearing in the eluate. The number of Na^+ ions in the eluate usually increases rapidly, and when the exchanger becomes fully saturated with these ions the NaCl solution passes unchanged through the bed. This process can be illustrated by curves representing the c/c_0 ratio (giving the ratio of eluate concentration c to input concentration c_0) as a function of time or volume of the eluate. Such curves called isoplanes or break-through curves are shown in Fig. 4.30, where the shaded area represents the total exchange capacity of the bed; for symmetrical break-through curves this capacity is equal to the abscissa b corresponding to the point $c/c_0=0.5$. Even when the experimental curves deviate from symmetry, section b can be used in most cases as the approximate value of the total capacity. Section a corresponds to the break-through volume of the bed under the given operating conditions.

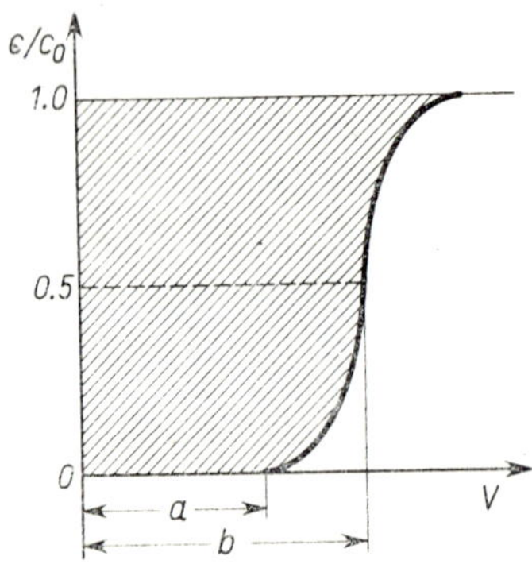

FIG. 4.30 Break-through isoplane (curve): section a is the break-through volume of the column under the given conditions, section b is the total exchange of the column.

4.4.4 Theory Underlying the Ion Exchange Process

Attempts have been made to find a suitable explanation of the ion exchange process on the basis of ample experimental data and theoretical considerations. So far, however, no fully developed theory of the mechanism of ion exchange exists. Since the ion exchange process is accompanied by a number of other processes such as adsorption, chemisorption, absorption and catalytic reactions it is easy to see that a one-sided consideration of the ion exchange process would be inaccurate and lead to an erroneous interpretation.

Different standpoints of the various authors who tried to explain the ion exchange process were due primarily to their use of different ion exchangers and different research techniques.

Ion exchange in inorganic exchangers has been explained by the exchange of ions within the crystal lattice. The exchange effect depends on the forces holding the ions in the lattice, the concentration of and charges on the ions in solution, the ionic radii, etc.

Sorption theories and resultant empirical isotherms have been used for the intepretation and qualitative analysis of the exchange process using appropriately modified Freundlich or Langmuir equations.

Ion exchangers have also been considered as gels, structurally similar to colloidal electrolytes, and quantitative aspects of the exchange equilibrium are based on the Donnan equilibrium theory. This theory is not inconsistent with the crystal lattice theory, but presents ion exchange phenomena in mathematical terms.

The Donnan theory has served as a basis for the swelling theory, proposed by Procter and Wilson [88], which treats the ion exchanger gel as an osmotic cell with an elastic film, and the distribution of ions between the solid phase and the solution is determined from the Donnan equation. An independent swelling theory has been proposed by Gregor [89] who treats ion exchangers as hydrated polyelectrolyte gels and assumes that their swelling is due to osmotic pressure. According to this theory, swelling of the gels occurs as a result of the hydration capacity of the functional groups of the ion exchanger and is limited by the elastic forces of the polymer. Since there exist macroions, attached to the polymer matrix, which are incapable of diffusion and different electrolyte concentrations occur in both phases (gel and liquid), an osmotic pressure difference is established between these phases which allows water and hydrated ions to penetrate into the gel phase.

4.4.5 Factors Affecting the Ion Exchange Process

Ion exchange depends as much on the properties of the exchanger as on the properties of the ions undergoing exchange. The affinity of an ion towards a given ion exchanger, i.e. the ion exchangeability, depends primarily on the electric charge of the ion, the ionic radius and degree of hydration. The larger the charge on the ion the greater is the force with which it is attracted by the functional groups of opposite charge on the ion exchanger, and hence the larger is its exchange capacity, i.e. the more difficult it is to remove during the exchanger-regeneration process.

In the case of equivalent ions, the magnitude of their radii is decisive for their exchange capacity. The greater the volume of the ion, the weaker is its electric field in the solution and thus the smaller is its degree of hydration. The so-called hydrodynamic radii of ions decrease with increasing atomic weight and hence their exchange energy (energy with which the ion is transported from the solution into the ion exchanger) increases.

We thus derive the rule that the exchange capacity of cations is inversely proportional to the hydrated ion radius. This rule does not hold, however, for ions of transition elements.

It should be noted that the degree of ion hydration depends on solution concentration, temperature, presence of contaminants and similar factors, and is therefore a variable quantity.

Ions can be arranged in a series according to their exchange energy. Such series have been established for cations and anions, where the order of ions in the given series naturally depends on the properties of the solution (pH, concentration, nature of solvent). The cation series for sulphonated phenolic resin ion exchangers is given below [90]:

$$Na^+ < NH_4^+ < K^+ < Sr^{2+} < Cs^+ < Mg^{2+} < Ca^{2+} < Cd^{2+} <$$
$$< Co^{2+} < Al^{3+} < Fe^{3+}.$$

For weakly basic anion exchangers we have the following series of anions [91]:

$$OH^- > SO_4^{2-} > CrO_4^{2-} > \text{citrate} > NO_3^- > PO_4^{3-} > \text{acetate} >$$
$$> I^- > Br^- > Cl^- > F^-.$$

The exchangeability of ions depends largely on the pH of the solution. Two phenomena are of paramount importance here; the degree of dissociation of the functional groups of the ion exchanger and the presence of H^+ and OH^- ions in the solution, competing with other ions being exchanged. The acidic or basic dissociation of organic electrolytes is dependent upon the solution pH.

The exchangeability of ions also depends on solution concentration, as follows from the equations describing the exchange equilibrium.

4.5 Thermodynamics of Adsorption from Solution

The thermal effects of wetting solid adsorbents with pure liquids and their solutions were studied as early as the thirties. Attempts to treat adsorption at the solid/solution interface in a systematic manner were described in post World War II publications. In the last two decades great interest has been shown in adsorption from solutions. Of fundamental importance are the works by Kiselev *et al.* [6, 7, 92], Everett [8, 9], Schay [93], Sircar,

Myers *et al.* [84, 94, 95], and Rusanov [22, 48]. Despite intensive studies, however, the provision of a thermodynamic interpretation of adsorption from solutions has not as yet been fully solved. The considerations below refer only to certain problems which lend themselves to experimental studies.

Kiselev [92] first considered the thermodynamics of adsorption from solutions in terms of the surface excess. He concluded that the total amount of adsorbed substance in the surface layer should be considered rather than the surface excess amounts. Eventually all adsorbate molecules interact with the adsorbent surface, not only their excess amount calculated with respect to the reference system (for which there is no adsorption).

Consider the variation of Gibbs free energy in the formation of the surface layer in adsorption from binary solutions at constant temperature and pressure. We shall constantly refer to the work of Kiselev and Pavlova published in 1962 [6].

Immersion of 1 g of adsorbent of specific surface area s in n_2^0 moles of substance 2 (solvent) is accompanied by a change of the Gibbs free energy of the system equal to $\Delta_w G_2^0$ ($\Delta_w G_2^0 = W_{w,2}^0$ is the work of wetting). The quantity $\Delta_w G_2^0$ is proportional to the specific surface area of adsorbent. Introduction of n_1^0 moles of substance 1 into the liquid 2/adsorbent system causes the displacement from the surface layer of a number of molecules of substance 2 by the adsorbed molecules of substance 1. At adsorption equilibrium we have:

in the adsorption layer: n_1^s moles of substance 1,
n_2^s moles of substance 2;
in the bulk phase: $n_1 = n_1^0 - n_1^s$ moles of substance 1,
$n_2 = n_2^0 - n_2^s$ moles of substance 2.

Assuming the symmetric reference system for the surface and bulk solutions the change of Gibbs free energy due to the above process is:

$$\Delta G_{1,2} = \int_0^{n_1^s} RT(\ln x_1 f_1 - \beta \ln x_2 f_2)\, dn_1^s + n_1 RT \ln x_1 f_1 + n_2 RT \ln x_2 f_2 \tag{4.103}$$

where x_1, x_2 and f_1, f_2 are mole fractions and activity coefficients, respectively, of the components in the bulk solution at adsorption equilibrium, and $\beta = n_{m,2}^s / n_{m,1}^s$ is the coefficient of surface displacement. The term $\Delta G_{1,2}$ refers to the total change of Gibbs free energy comprising two contributions:

(i) due to displacement from the surface layer of $(n_{m,2}^s - n_2^s)$ moles of substance 2 by n_1^s moles of substance 1,

(ii) due to mixing of n_1 moles of substance 1 with n_2 moles of substance 2.

The change of Gibbs free energy due to the mixing of the solution components is:

$$\Delta G_l = n_1 RT \ln x_1 f_1 + n_2 RT \ln x_2 f_2 . \tag{4.104}$$

Comparing equations (4.103) and (4.104) we find that the change of Gibbs free energy due to displacement from the surface layer of molecules of substance 2 by molecules of substance 1, i.e. the Gibbs free energy of adsorption (or work of adsorption) from solution, is:

$$\Delta_a G_1^s = \int_0^{n_1^s} RT(\ln x_1 f_1 - \beta \ln x_2 f_2)\, dn_1^s . \tag{4.105}$$

We can also calculate the differential molar Gibbs free energy of adsorption:

$$\Delta_a G_{m,1}^s = \left(\frac{\partial \Delta_a G_1^s}{\partial n_1^s}\right)_{T,p} = RT(\ln x_1 f_1 - \beta \ln x_2 f_2)_{n_1^s} . \tag{4.106}$$

For an ideal solution (bulk phase) we can write equation (4.106) in the form:

$$\Delta_a G_{m,1}^s = RT(\ln x_1 - \beta \ln x_2)_{n_1^s} = RT\left[\ln \frac{x_1}{(1-x_1)^\beta}\right]_{n_1^s} . \tag{4.107}$$

The differential molar Gibbs free adsorption energy is equivalent to the work necessary to introduce 1 mole of substance 1 into the surface layer at equilibrium. This occurs with the simultaneous displacement of β moles of substance 2 from that layer. Since the process is carried out at equilibrium (virtual displacement), neither the equilibrium concentration of the bulk solution nor the composition of the surface layer may change. We can imagine this in a system with a large volume of solution and large surface area of adsorbent (for an adequately large mass of adsorbent).

We assume above that the adsorbent was initially immersed in liquid substance 2 and subsequently substance 1 was added. However, if we immerse 1 g of adsorbent directly in a solution with initial concentrations x_1^0 and x_2^0 of the components containing n_1^0 and n_2^0 moles of these components, then so-called Gibbs free energy of wetting of the adsorbent with the solution is equal to the sum:

$$\Delta_w G = \Delta_w G_2^0 + \Delta_a G_1^s + \Delta G_l - \Delta G_l^0 \tag{4.108}$$

where $\Delta_w G_2^0$ is the Gibbs free energy of adsorbent wetting with pure component 2, and ΔG_l^0 is the change of the Gibbs free energy of the system related to formation of the initial solution from the pure components 1 and 2:

$$\Delta G_1^0 = n_1^0 RT \ln x_1^0 f_1^0 + n_2^0 RT \ln x_2^0 f_2^0. \tag{4.109}$$

Equation (4.108) is obvious, since prior to the above adsorption process we can always separate the initial solution into pure components performing the work $W_1^0 = \Delta G_1^0$. In accordance with the Gibbs–Helmholtz equation we can write:

$$\Delta_a H_1^s = \Delta_a G_1^s - T\left(\frac{\partial \Delta_a G_1^s}{\partial T}\right)_{n_1^s} \tag{4.110}$$

where $\Delta_a H_1^s$ is the enthalpy change arising from the displacement of $(n_{m,2}^s - n_2^s)$ moles of substance 2 from the surface layer by n_1^s moles of substance 1, and thus it is the so-called enthalpy of adsorption (heat of adsorption) from solution. Assuming that the coefficient β changes only slightly with temperature and substituting equation (4.105) into equation (4.110) we get:

$$\Delta_a H_1^s = RT^2 \int_0^{n_1^s} \left[\beta\left(\frac{\partial \ln x_2 f_2}{\partial T}\right)_{n_1^s} - \left(\frac{\partial \ln x_1 f_1}{\partial T}\right)_{n_1^s}\right] dn_1^s \tag{4.111}$$

or

$$\Delta_a H_1^s = R \int_0^{n_1^s} \left[\frac{\partial}{\partial(1/T)} (\ln x_1 f_1 - \beta \ln x_2 f_2)\right]_{n_1^s} dn_1^s. \tag{4.112}$$

The differential molar enthalpy of adsorption from solution (differential molar heat of adsorption) $\Delta_a H_{m,1}^s$ can also be calculated as follows:

$$\Delta_a H_{m,1}^s = \left(\frac{\partial \Delta_a H_1^s}{\partial n_1^s}\right)_{T,p} = R\left[\frac{\partial}{\partial(1/T)} (\ln x_1 f_1 - \beta \ln x_2 f_2)\right]_{n_1^s}. \tag{4.113}$$

For solutions close to ideality, when $f_1 = f_2 = 1$, we can write:

$$\Delta_a H_{m,1}^s = R\left[\frac{\partial}{\partial(1/T)} (\ln x_1 - \beta \ln x_2)\right]_{n_1^s} = R\left[\frac{\partial}{\partial(1/T)} \ln \frac{x_1}{(1-x_1)^\beta}\right]_{n_1^s}. \tag{4.114}$$

Using equation (4.113) or (4.114) we can calculate the differential molar enthalpy of adsorption for a binary solution from the adsorption isosteres or individual adsorption isotherms $n_1^s = f(x_1)$ measured at different temperatures. This quantity is therefore analogous to the isosteric heat of adsorption of a gas (vapour) on a solid (of course $\Delta_a H_{m,1}^s = -q_{st}$, cf. Chapter 3).

NOTE. In the case of adsorption from solution the effect of temperature is much smaller than for adsorption from the gaseous phase. It is necessary therefore to measure the adsorption isotherm at temperatures which differ for example, by about 20 to 30 deg.

The term $\Delta_a H^s_{m,1}$ is the enthalpy change related to the virtual displacement of 1 mole of substance 1 from the solution to the adsorption layer accompanied by the displacement from this layer to the bulk solution of β moles of substance 2. At equilibrium, of course, neither the composition of the adsorption layer nor that of the equilibrium solution may change.

If we measure, using a calorimeter, the enthalpy of adsorption of component 1 by adding it to 1 g of adsorbent previously wetted with component 2, then we get an enthalpy change given by the sum:

$$\Delta H_{1,2} = \Delta_a H^s_1 + \Delta H_l \tag{4.115}$$

where ΔH_l is the enthalpy of formation of the equilibrium solution from its component (heat of solution formation).

The enthalpy change connected with the immersion of 1 g of adsorbent in the solution, i.e. the so-called enthalpy (or heat) of wetting with the solution is:

$$\Delta_w H = \Delta_w H^0_2 + \Delta_a H^s_1 + \Delta H_l - \Delta H^0_l \tag{4.116}$$

where $\Delta_w H^0_2$ is the enthalpy of wetting of 1 g of adsorbent with pure component 2, ΔH_l and ΔH^0_l are the enthalpies of formation of the equilibrium and initial solutions, respectively (the latter quantities are the enthalpies or heats of mixing). The difference $\Delta H_l - \Delta H^0_l$, the enthalpy of dilution, may have a significant effect on the isotherms of wetting of the adsorbent with the solutions.

Kiselev and Pavlova [96] have measured excess adsorption isotherms of benzene from *n*-hexane on silica gels *SG* and SDG-2 (silica gel with dehydrated surface) at two different temperatures: 253 and 293 K. They found from experimental data the individual adsorption isotherms of benzene $\alpha^s_{C_6H_6} = f(x_{C_6H_6})$ (with $\alpha^s_{C_6H_6} = n^s_{C_6H_6}/s$), and subsequently, using equation (4.106) or (4.107), they calculated the Gibbs free energy of adsorption of benzene from *n*-hexane at 293 K. Their results are given in Fig. 4.31 where the $\Delta_a G_{m,\,C_6H_6}$ vs. $\alpha^s_{C_6H_6}$ relation is plotted. The isosteric differential molar heats of adsorption of benzene from *n*-hexane on commonly used silica gels were calculated using equation (4.113) or (4.114). The $q_{st\,C_6H_6}$ vs. $\alpha^s_{C_6H_6}$ plot is shown in Fig. 4.32, where it is seen that the differential molar heat of adsorption from solution only varies at very small values of adsorption; at higher values it remains almost constant.

Laryonov *et al.* [97, 98] have studied adsorption of carbon tetrachloride and of benzene from iso-octane on activated carbon at 303 and 338 K. Making use of the R–D equation they determined [49, 53] from the adsorption of vapours at the given temperatures (with $\theta=1$) the values of $n^s_{m,i}$ (getting $n^s_{m,CCl_4}=3.61$ and 3.45 mmole/g, $n^s_{m,\text{iso-octane}}=2.15$ and 2.08 mmole/g, at the two temperatures respectively and $n^s_{m,C_6H_6}=4.315$ at both

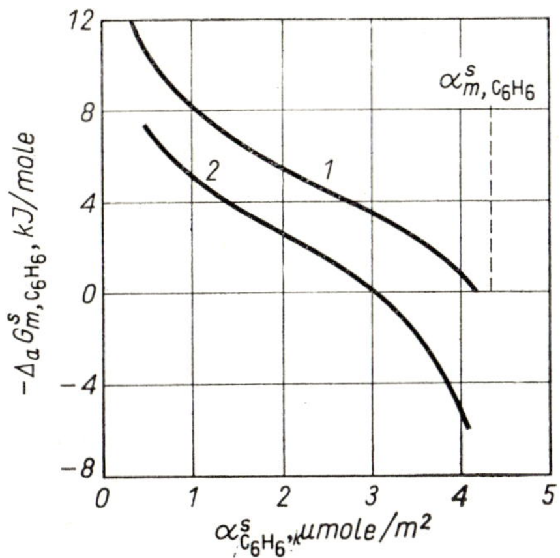

FIG. 4.31 Variations of the molar differential Gibbs free energy of benzene adsorption ($-\Delta_a G^s_{m,C_6H_6}$) from *n*-hexane on silica gels with surface concentration $\alpha^s_{C_6H_6}$: for silica gels SG (*1*) and SDG-2 (*2*) (after Kiselev and Pavlova [96]).

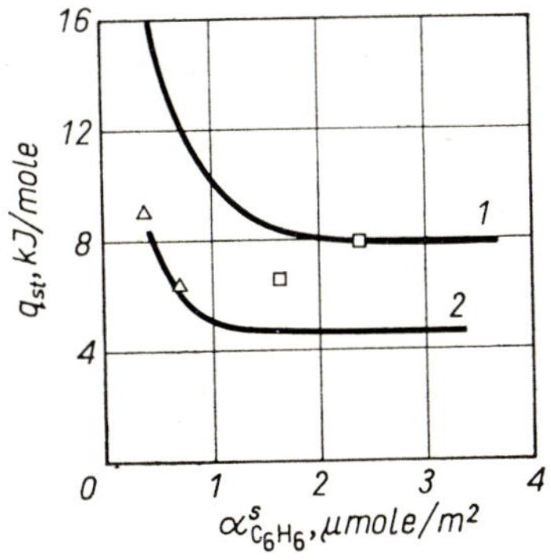

FIG. 4.32 Variation of the isosteric differential molar heat of adsorption of benzene ($q_{st\,C_6H_6}$) on silica gel with surface concentration $\alpha^s_{C_6H_6}$: for silica gels SG (*1*) and SDG-2 (*2*) (after Kiselev and Pavlova [96]).

temperatures). Their experimental results allowed them to study the variations of $\Delta_a G^s_{m,i}$, $\Delta_a H^s_{m,i}$ and $T\Delta_a S^s_{m,i}$ for carbon tetrachloride and benzene adsorbed from iso-octane with variation of adsorbent coverage θ (Figs. 4.33 and 4.34).

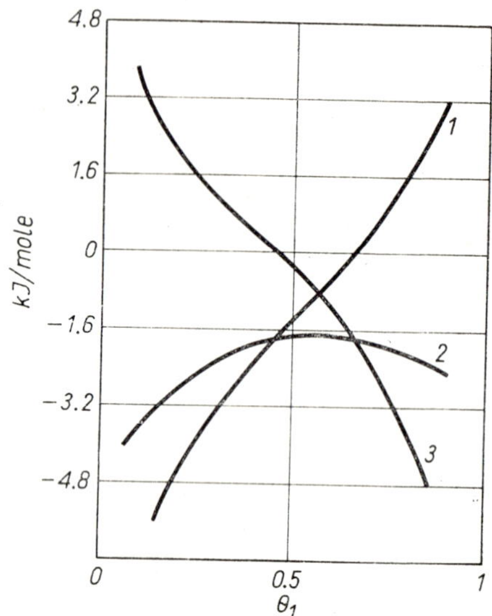

FIG. 4.33 Variation of the differential thermodynamic functions of CCl_4 adsorption from iso-octane on activated carbon with adsorbent surface coverage θ_1: *1*—Gibbs free energy ($\Delta_a G^s_{m,CCl_4}$), *2*—enthalpy ($\Delta_a H^s_{m,CCl_4}$), *3*—entropy ($T\Delta_a S^s_{m\ CCl_4}$) (after Laryonov *et al.* [97, 98]).

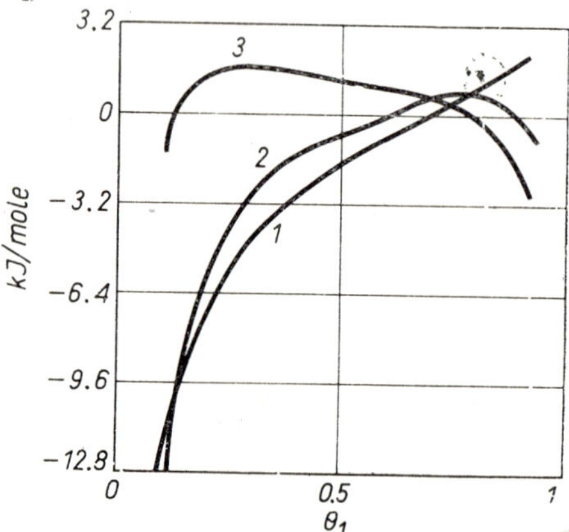

Fig. 4.34 Variation of the differential thermodynamic functions of benzene adsorption from iso-octane on activated carbon with adsorbent surface coverage θ_1: *1*—Gibbs free energy, ($\Delta_a G^s_{m,C_6H_6}$), *2*—enthalpy ($\Delta_a H^s_{m,C_6H_6}$), *3*—entropy ($T\Delta_a S^s_{m,C_6H_6}$) (after Laryonov *et al.* [97, 98]).

Everett [8] analysed the wetting enthalpy (heat of wetting or heat of immersion) of a uniform surface of 1 g of adsorbent with an ideal solution:

$$\Delta_w H = x_1^s \Delta_w H_1^0 + x_2^s \Delta_w H_2^0 \tag{4.117}$$

where $\Delta_w H_1^0$ and $\Delta_w H_2^0$ are the enthalpies of wetting of the surface of 1 g of adsorbent with pure liquids 1 and 2; these quantities (like $\Delta_w G_1^0$ and $\Delta_w G_2^0$) can also be related to 1 mole of substance, when they represent the difference of the enthalpies of 1 mole of substance in the surface layer and in the liquid bulk phase. We can then use the notation: $\Delta_w H_{m,1}^0$ and $\Delta_w H_{m,2}^0$. The quantity $\Delta_w H$ is then, according to equation (4.117), the difference of the enthalpy of an average mole of the solution in the surface layer and the same average mole of solution in the liquid bulk phase. This difference can be denoted as $\Delta_w H_m$.

Considering equation (4.19) we can write:

$$x_1^s - x_1 = \frac{x_1 x_2}{x_1 + (K_1 - 1)^{-1}} . \tag{4.118}$$

From equations (4.117) and (4.118) we get:

$$\Delta_w H - (x_1 \Delta_w H_1^0 + x_2 \Delta_w H_2^0) = \frac{x_1 x_2}{x_1 + (K_1 - 1)^{-1}} (\Delta_w H_1^0 - \Delta_w H_2^0) . \tag{4.119}$$

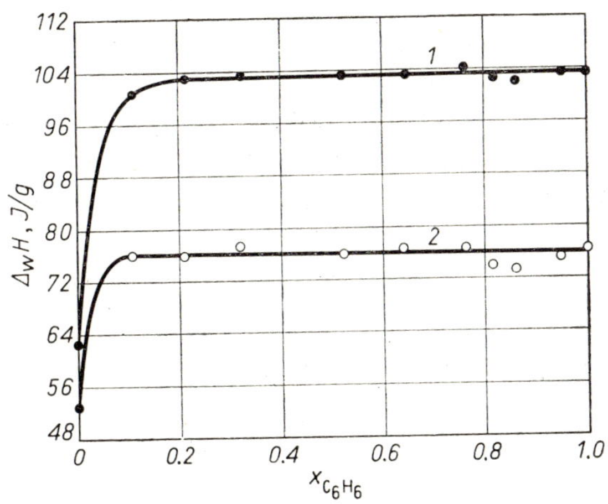

Fig. 4.35 Variation of the heats of wetting of activated carbons with benzene concentration in cyclohexane: *1*—Shell carbon, *2*—decolourizing carbon (after Wright [100]).

From a knowledge of the measured values of the enthalpy of wetting, $\Delta_w H$, $\Delta_w H_1^0$ and $\Delta_w H_2^0$, we can investigate, using the latter equation, the thermodynamical validity of the Everett model of adsorption from solution. The $(\Delta_w H - x_1 \Delta_w H_1^0 - x_2 \Delta_w H_2^0)$ vs. $x_1 x_2/[x_1 + (K_1 - 1)^{-1}]$ plot should be linear with a gradient of $\Delta_w H_1^0 - \Delta_w H_2^0$.

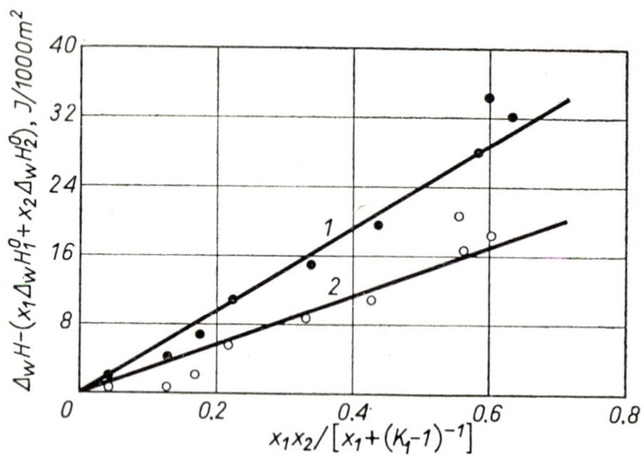

Fig. 4.36 Heats of wetting of activated carbons with benzene–cyclohexane solution in the co-ordinate system from equation (4.119): *1*–Shell carbon, *2*–decolourizing carbon (after Wright [100]).

Equation (4.119) was applied by Wright [99, 100] in studies of adsorption from binary solutions. His results for adsorption of benzene from cyclohexane on Shell and decolourizing activated carbons are presented in Figs. 4.35 and 4.36 and in Table 4.3. Good agreement is obtained for these systems between experimental results and theoretical predictions.

Table 4.3 THEORETICAL AND EXPERIMENTAL DATA OF HEATS OF WETTING AFTER WRIGHT [100]

Adsorption system	$\Delta_w H_1^0 - \Delta_w H_2^0$ calculated from equation (4.119) J/g	$\Delta_w H_1^0 - \Delta_w H_2^0$ found experimentally J/g
Benzene + cyclohexane Shell carbon	50.208	42.89 ± 6.28
Benzene + cyclohexane Decolourizing carbon	29.874	24.77 ± 6.28

Everett [9] has also derived an equation for the wetting (immersion) enthalpy for a non-ideal solution. We write:

$$\Delta_w H-(x_1 \Delta_w H_1^0+x_2 \Delta_w H_2^0)=\Delta x_1(\Delta_w H_1^0-\Delta_w H_2^0)+ \\ +\Delta H_l+\Delta H_l^s-\Delta H_l^0 \qquad (4.120)$$

where $\Delta x_1=x_1^0-x_1(x_1^0$ is the mole fraction of component 1 in the initial solution), and ΔH_l^s is the enthalpy of formation of the surface solution on 1 g of adsorbent. Equation (4.120) can be written as:

$$\Delta_w H-(x_1^s \Delta_w H_1^0+x_2^s \Delta_w H_2^0)=\Delta H_l^s+\Delta H_l-\Delta H_l^0. \qquad (4.121)$$

The right-hand side is equal to zero for an ideal solution, so for a non-ideal solution it represents the excess heat of wetting (immersion). Lu and Lama [101] have investigated this excess for the case of immersion of three types of silica gels in benzene–cyclohexane binary mixtures at 298 K. They measured simultaneously adsorption of benzene from cyclohexane on the various adsorbents, calculating the plot of the $x_{C_6H_6}^s=f(x_{C_6H_6})$ function. Their results are given in Figs. 4.37 and 4.38. The specific surface areas of the silica

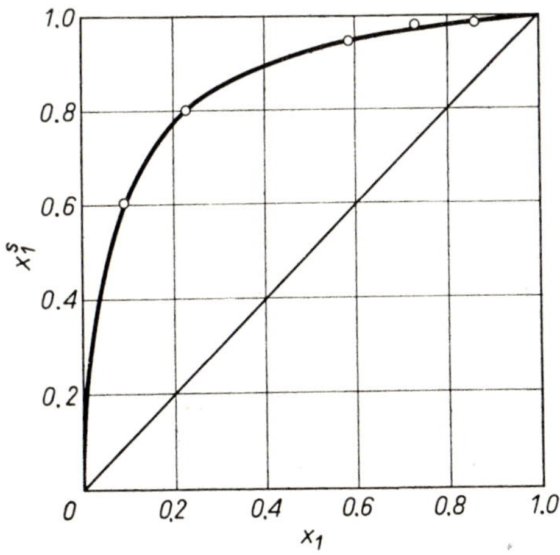

Fig. 4.37 Individual benzene adsorption isotherm from cyclohexane on silica gel I: $x_{C_6H_6}^s=f(x_{C_6H_6})$ (after Lu and Lama [101]).

gels used were: I: 666, II: 575, III: 483 m²/g. The excess wetting heats were calculated per 1000 m² of the adsorbent surface area. Making use of the values of ΔH_l and ΔH_l^0, measured by Lundberg [102], it was possible to calculate H_l^s which proved to be much larger than ΔH_l (for the same composition of the surface and bulk solution).

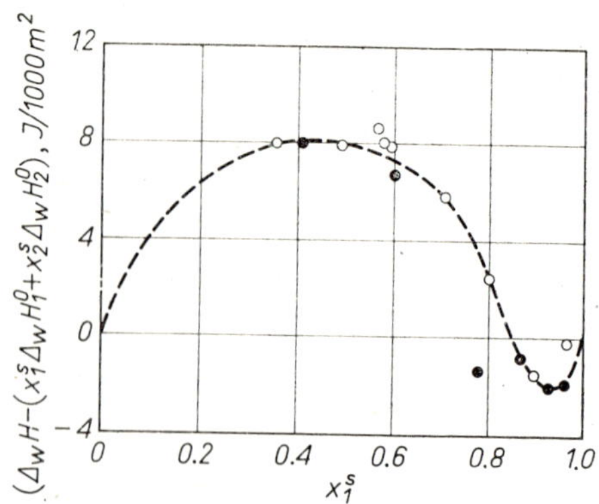

Fig. 4.38 The excess heat of wetting (immersion) as a function of the mole fraction of benzene ($x_{C_6H_6}$) in cyclohexane: open circles – silica gel II, full circles – silica gel III (after Lu and Lama [101]).

NOTE. It is interesting to compare the equations describing the enthalpy of wetting of adsorbent with solution derived by Kiselev and Pavlova [6] (equation (4.116)) and by Everett [9] (equation (4.121)). We then get:

$$\Delta_w H_2^0+\Delta_a H_1^s=x_1^s\,\Delta_w H_1^0+x_2^s\,\Delta_w H_2^0+\Delta H_1^s. \tag{4.122}$$

After transformation we can write equation (4.122) in the form:

$$\Delta_a H_1^s=x_1^s(\Delta_w H_1^0-\Delta_w H_2^0)+\Delta H_1^s \tag{4.123}$$

Studies of adsorption from solution are often conducted at low concentrations (for the initial sections of the adsorption isotherms). The differential molar Gibbs free energy of adsorption of component i from the solution can then be calculated [102–105] from the equation:

$$\Delta_a G_{m,i}^s=-RT\ln\frac{x_i^s}{x_i}. \tag{4.124}$$

Assuming an infinitely dilute solution as standard state we obtain an expression for the standard differential molar free enthalpy of adsorption of component i from solution which is a measure of its adsorption affinity in the given adsorbent/solvent system [25, 27]:

$$\Delta_a G_{m,i}^{\ominus,s}=\Delta\mu_i^{\ominus}=-RT\lim_{x_i\to 0}\ln\frac{x_i^s}{x_i}. \tag{4.125}$$

For calculating $\Delta_a G_{m,i}^{\ominus,s}$ we can also use the constant k from the Langmuir adsorption isotherm in which the pressure p_i is replaced by the concentration x_i, provided the initial section of the individual adsorption isotherm $x_i^s = f(x_i)$ can be described by that equation [106].

Wright and Powell [105] calculated using, for instance, equation (4.124), values of $\Delta_a G^s_{m,i}$ for some aromatic hydrocarbons adsorbed from cyclohexane on activated carbons Graphon and Spheron 6. The concentrations (in mole fractions) of the aromatic hydrocarbons in equilibrium solutions did not exceed 0.15. These authors also assumed that for dilute solutions:

$$\Delta_a H^s_{m,i} = x_i^s \Delta_w H^0_{m,i} \tag{4.126}$$

where $\Delta_w H^0_{m,i}$ is the molar enthalpy of wetting with the pure component i. Making use of the Gibbs–Helmholtz equation we can calculate $\Delta_a S^s_{m,i}$, viz.:

$$\Delta_a S^s_{m,i} = \frac{1}{T} (\Delta_a H^s_{m,i} - \Delta_a G^s_{m,i}) . \tag{4.127}$$

The method of Wright and Powell may, in many instances, be useful for estimating molecular interactions in the process of adsorption from solution.

4.6 Kinetics of Adsorption from Solution

The rate of adsorption from solution depends on many factors. Of primary importance here are: size and structure of the adsorbed molecules, nature of solvent and porosity of the adsorbent.

Every adsorption process (like every heterogeneous process) has two successive steps.

(i) Transport of adsorbate from the bulk phase to the adsorbent surface.

(ii) Adsorption on the adsorbent surface.

The first step depends on the nature of solvent and adsorbate, and is controlled by the laws of diffusion. Pouchlý and Erdös, when considering the kinetics of adsorption from solution, concluded [107] that in this step one should distinguish:

– diffusion of adsorbate to the external surface of the adsorbent (external diffusion),

– diffusion of adsorbate into the adsorbent capillaries (internal diffusion).

According to these authors, the process of adsorption itself is so rapid that it is difficult to determine its time of duration.

Experimental data indicate that both in adsorption from the gaseous phase and in adsorption from solution, most of the adsorbed substance passes to the adsorbent surface layer in a relatively short time. Adsorption equilibrium, however, takes longer to establish.

References

[1] Manes, M. and Hofer, L. J. E., *J. Phys. Chem.*, **73**, 584 (1969).
[2] Wohleber, D. A. and Manes, M., *J. Phys. Chem.*, **75**, 61 (1971).
[3] Wohleber, D. A. and Manes, M., *J. Phys. Chem.*, **75**, 3720 (1971).
[4] Kiselev, A. V., *Usp. Khim.*, **15**, 456 (1946).
[5] Dzhygit, O. M., Kiselev, A. V., Terekhova, M. G. and Shcherbakova, K. O., *Zh. Fiz. Khim.*, **22**, 107 (1948).
[6] Kiselev, A. V. and Pavlova, L. F., *Izv. Akad. Nauk SSSR, Otd. Khim. Nauk*, No. 12, 2121 (1962).
[7] Kiselev, A. V. and Chopina, V. V., *Trans. Faraday Soc.*, **65**, 1936 (1969).
[8] Everett, D. H., *Trans. Faraday Soc.*, **60**, 1803 (1964).
[9] Everett, D. H., *Trans. Faraday Soc.*, **61**, 2478 (1965).
[10] Kipling, J. J., *Adsorption from Solution of Non-electrolytes*, Academic Press, London–New York 1965.
[11] Schay, G., Nagy, L. and Szekrenyesy, T., *Period. Polytechnica*, **4**, 95 (1960).
[12] Schay, G. and Nagy, L., *J. chim. phys.*, **58**, 149 (1961).
[13] Schay, G. and Nagy, L., *Period. Polytechnica*, **6**, 91 (1962).
[14] Jelovich, S. and Laryonov, O. G., *Izv. Akad. Nauk SSSR, Otd. Khim. Nauk*, No. 2, 209 (1962).
[15] Laryonov, O. G., Tonkong, L. G. and Chmutov, K. V., *Zh. Fiz. Khim.*, **39**, 2226 (1965).
[16] Laryonov, O. G., *Zh. Fiz. Khim.*, **40**, 1796 (1966).
[17] Laryonov, O. G., Chmutov, K. V. and Yubilevich, M. D., *Zh. Fiz. Khim.*, **41**, 2616 (1967).
[18] Šiškova, M. and Erdös, E., *Coll. Czechoslov. Chem. Comm.*, **25**, 1729 (1960).
[19] Šiškova, M. and Erdös, E., *Coll. Czechoslov. Chem. Comm.*, **25**, 2599 (1960).
[20] Šiškova, M., Erdös, E. and Kadlec, O., *Coll. Czechoslov. Chem. Comm.*, **39**, 1954 (1974).
[21] Rusanov, A. I., *Termodinamika poverkhnostnykh yavlenii* (*Thermodynamics of Surface Phenomena*), Izd. Leningradskogo Univ., Leningrad 1960.
[22] Rusanov, A. I., in: *Progress Surface and Membrane Sci.*, Vol. 4, p. 57, Academic Press Inc., New York–London 1971.
[23] Fu, Y., Hansen, R. S. and Bartell, F. E., *J. Phys. Chem.*, **52**, 374 (1948).
[24] Elton, G. A. H., *J. Chem. Soc.*, 2958 (1954).
[25] Ościk, J. and Waksmundzki, A., *Ann. Univ. Maria Curie-Skłodowska, Lublin, AA*, **9**, 9 (1954).
[26] Ościk, J., *Ann. UMCS, Lublin, AA*, **19**, 15 (1964).
[27] Ościk, J., *Przemysł Chem.*, **44**, 129 (1965).

[28] Malesiński, W., *Termodynamiczne własności roztworów ciekłych* (*Thermodynamic Properties of Liquid Solutions*) in: *Teoria roztworów ciekłych w stanie równowagi* (*Theory of Liquid Solutions in Equilibrium*), Biuro Kształcenia i Doskonalenia Kadr Naukowych, PAN, Warszawa 1961.
[29] Butler, J. A. V., *Proc. Roy. Soc.*, **135**, 348 (1932).
[30] Kiselev, A. V. and Chopina, V. V., *Trans. Faraday Soc.*, **65**, 1936 (1969).
[31] Ostwald, W. and Izaguirre, R., *Kolloid Z.*, **30**, 279 (1922).
[32] Nagy, L. G. and Schay, G., *Magyar Kem. Folyoirat.*, **60**, 31 (1960).
[33] Schay, G., *Acta Chim. Acad. Sci. Hung.*, **10**, 281 (1956).
[34] Dzhygit, O. M., Kiselev, A. V. and Krasilnikov, K. G., *Dokl. Akad. Nauk SSSR*, **58**, 413 (1947).
[35] Kiselev, A. V. and Kulichenko, V. V., *Dokl. Akad. Nauk SSSR*, **82**, 89 (1952).
[36] Krasilnikov, K. G. and Kiselev, A. V., *Dokl. Akad. Nauk SSSR*, **63**, 693 (1948).
[37] Giles, C. H., MacEvan, T. H., Nakhawa, S. W. and Smith, D., *J. Chem. Soc.*, 3973 (1960).
[38] Giles, C. H. and Smith, D., *J. Colloid and Interface Sci.*, **47**, 755 (1974).
[39] Giles, C. H., D'Silva, A. P. and Baston, J. A., *J. Colloid and Interface Sci.*, **47**, 766 (1974).
[40] Lipatov, Yu. S. and Sergeeva, L. M., *Adsorbtsiya polimerov* (*Adsorption of Polymers*), Izd. Naukova Dumka, Kiev 1972.
[41] Kurbanbekov, E., Laryonov, O. G., Chmutov, K. V. and Yubilevich, M. D., *Zh. Fiz. Khim.*, **43**, 1630 (1969).
[42] Kiselev, A. V. and Yashin, Yu. I., *Gazo-adsorbtsionnaya khromatografiya* (*Adsorption Gas Chromatography*), Izd. Nauka, Moscow 1967.
[43] Hadden, N., Baumann, F., Macdonald, F., Mink, M., Stevenson R., Gere, D. and Zamaroni, F., *Basic Liquid Chromatography*, Varian (USA) 1971.
[44] Kiselev, A. V. and Shcherbakova, K. D., *Acta Physicochem. U.S.S.R.*, **21**, 539 (1946).
[45] Kiselev, A. V. and Eltekov, Yu. A., *Dokl. Akad. Nauk SSSR*, **100**, 108 (1955).
[46] Schay, G. and Nagy, L. G., *J. Colloid and Interface Sci.*, **38**, 302 (1972).
[47] Serpinski, V. V., *Fizicheskaya adsorbtsya iz mnogokomponentnykh faz* (*Physical Adsorption from Multi-component Phases*), Izd. Nauka, Moscow 1972, p. 124.
[48] Rusanov, A. I., *Fazovye ravnovesiya i poverkhnostnye yavlenya* (*Phase Equilibria and Surface Phenomena*), Izd. Khimia, Leningrad 1967.
[49] Bering, B. P. and Serpinski, V. V., *Izv. Akad. Nauk SSSR, Ser. Khim.*, No. 6, 1232 (1970).
[50] Schay, G., *Surface Area Determination*, IUPAC and Soc. Chem. Int., Proc. International Symp. 1969, Butterworths, London 1970, p. 273.
[51] Blackburn, A. and Kipling, J. J., *J. Chem. Soc.*, 3819 (1954).
[52] Kipling, J. J. and Peakall, D. B., *J. Chem. Soc.*, 4828 (1956).
[53] Laryonov, O. G. and Kurbanbekov, E., *Fizicheskaya adsorbstsya iz mnogokomponentnykh faz* (*Physical Adsorption from Multi-c mponent Phases*), Izd. Nauka, Moscow 1972, p. 85.
[54] Klinkenberg, A., *Rec. trav. chim.*, **78**, 83 (1959).
[55] Ościk, J., Dąbrowski, A. and Rudziński, W., *Materiały IV Kolokwium "Adsorpcja i adsorbenty"* (*Proceedings of the 4th Conference on Adsorption and Adsorbents*), 1974, Inst. Energochemii Węgla i Fizykochemii Sorbentów, AGH, Kraków.

[56] Ościk, J., Dąbrowski, A., Jaroniec, M. and Sokołowski, S., *J. Catalysis, Hokkaido Univ., Japan*, **23**, 91 (1975).
[57] Ościk, J., Dąbrowski, A., Jaroniec, M. and Rudziński, W., *J. Colloid and Interface Sci.*, **56**, 403 (1976).
[58] Ościk, J., Rudziński, W. and Dąbrowski, A., *Colloid and Polymer Sci.*, **255**, 50 (1977).
[59] Kiselev, A. V., Pavlova, L. F., *Kinetika i kataliz*, **2**, 599 (1961).
[60] Zhdanov, S. P., Kiselev, A. V. and Pavlova, L. F., *Kinetika i kataliz*, **3**, 445 (1962).
[61] Kipling, J. J. and Wright, E. H. M., *J. Chem. Soc.*, 3382 (1963).
[62] Kipling, J. J. and Wright, E. H. M., *J. Chem. Soc.*, 3535 (1964).
[63] Harkins, W. D. and Gans, D. M., *J. Am. Chem. Soc.*, **53**, 2804 (1931); *J. Phys. Chem.*, **36**, 86 (1932).
[64] Russel, A. S. and Cochran, C. N., *Ind. Eng. Chem.*, **42**, 1332 (1950).
[65] Daniel, S. G., *Trans. Faraday Soc.*, **47**, 1345 (1951).
[66] Gryazev, N. N., *Dokl. Akad. Nauk SSSR*, **118**, 317 (1958).
[67] Ościk, J., *Bull. Acad. Polon. Sci., sér. sci. chim.*, **9**, 29 (1961).
[68] Ościk, J., *Bull. Acad. Polon. Sci., sér. sci. chim.*, **9**, 33 (1961).
[69] Ościk, J., *Przemysł Chem.*, **40**, 281 (1961).
[70] Vasilyeva, V. S., Davydov, V. I. and Kiselev, A. V., *Dokl. Akad. Nauk SSSR*, **192**, 1299 (1970).
[71] Davydov, V. I., Kiselev, A. V. and Shtefans, D., *Zh. Fiz. Khim.*, **47**, 2091 (1973).
[72] Ościk, J., Kusak, R. and Goworek, J., *Materiały IV Kolokwium "Adsorpcja i adsorbenty"* (*Proceedings of the 4th Conference on Adsorption and Adsorbents*), 1974, Inst. Energochemii Węgla i Fizykochemii Sorbentów, AGH, Kraków.
[73] Minka, Ch. and Myers, A. I., *AIChE Journal*, **19**, 453 (1973).
[74] Schuchowitzky, A.. *Acta Physicochem. U.R.S.S.*, **8**, 531 (1938).
[75] Coltharp, M. T. and Hackerman, N., *J. Colloid Sci.*, **43**, 176 (1973); **43**, 185 (1973).
[76] Kagiya, T., Sumida, Y. and Tachi, T., *Bull. Chem. Soc. Japan*, **45**, 6 (1972).
[77] Kagiya, T., Sumida, Y. and Tachi, T., *Bull. Chem. Soc. Japan*, **44**, 1219 (1971).
[78] Rudziński, W., Ościk, J. and Dąbrowski, A., *Chem. Phys. Letters*, **20**, 5 (1973).
[79] Ościk, J., Rudziński, W. and Dąbrowski, A., *Roczniki Chem.*, **48**, 1991 (1974).
[80] Rudziński, W., Ościk, J. and Dąbrowski, A., *Adsorbtsya i poristost'* (*Adsorption and Porosity*), Izd. Nauka, Moscow 1976, p. 158.
[81] Ościk, J., Dąbrowski, A. and Jaroniec, M., *Physical Adsorption, 2nd Czechoslovak Conference, Liblice, 1975*, Czech. Akad. Nauk, Prague, p. 92.
[82] Toth, J., Rudziński, W., Ościk, J. and Dąbrowski, A., *Roczniki Chem.*, **48**, 1769 (1974).
[83] *Tables of Integral Transforms*, A. Erdelyi (Ed.), McGraw-Hill, New York 1954.
[84] Sircar, S. and Myers, A. L., *J. Phys. Chem.*, **74**, 2828 (1970).
[85] Sams, J. R., *Progr. Surface and Membrane Sci.*, **8**, 1 (1974).
[86] Radke, C. J. and Prausnitz, J. M., *J. Chem. Phys.*, **57**, 714 (1972).
[87] Stach, H. and Teichmüller, M., *Brennstoffchemie*, **34**, 275 (1953).
[88] Procter, H. R. and Wilson, J. A., *J. Chem. Soc.*, **109**, 307 (1916).
[89] Gregor, H. P., *J. Am. Chem. Soc.*, **70**, 1293 (1948); **73**, 642 (1951).
[90] Prokhorov, T. G. and Yankovski, K. A., *Zav. Lab.*, **13**, 658 (1947).

[91] Kunin, R. and Myers, R. J., *J. Am. Chem. Soc.*, **69**, 2874 (1947).
[92] Kiselev, A. V., *Dokl. Akad. Nauk SSSR*, **49**, 516 (1945); *Zh. Fiz. Khim.*, **20**, 239 (1946); *Usp. Khim.*, **25**, 705 (1956).
[93] Schay, G., *J. Colloid and Interface Sci.*, **42**, 478 (1973).
[94] Sircar, S., Nowosad, J. and Myers, A. L., *Ind. Eng. Chem. Fundam.*, **11**, 249 (1972).
[95] Laryonov, O. G. and Myers, A. L., *Chem. Eng. Sci.*, **26**, 1025 (1971).
[96] Kiselev, A. V. and Pavlova, L. F., *Neftkhimiya*, **2**, 861 (1962).
[97] Laryonov, O. G., Popov, E. A. and Chmutov, K. V., *Zh. Fiz. Khim.*, **47**, 1939 (1974).
[98] Laryonov, O. G., Popov, E. A. and Chmutov, K. V., *Zh. Fiz. Khim.*, **47**, 1843 (1974).
[99] Wright, E. H. M., *Trans. Faraday Soc.*, **62**, 1275 (1966).
[100] Wright, E. H. M., *Trans. Faraday Soc.*, **63**, 3026 (1967).
[101] Lu, B. C-Y. and Lama, R. F., *Trans. Faraday Soc.*, **63**, 727 (1967).
[102] Lundberg, G. W., *J. Chem. Eng. Data*, **9**, 193 (1964).
[103] Groszek, A. J., *J. Chromatography*, **3**, 454 (1960).
[104] Bartell, F. E. and Fu, Y., *J. Phys. Chem.*, **33**, 1759 (1929).
[105] Wright, E. H. M. and Powell, A. V., *Trans. Faraday Soc.*, **68**, 1908 (1972).
[106] Groszek, A. J., *Proc. Roy. Soc.* (*London*), **A 314**, 473 (1970).
[107] Pouchlý, J. and Erdös, E., *Coll. Czechoslov. Chem. Comm.*, **23**, 1706 (1958).

Chapter 5

Adsorption at the Liquid/Liquid Interface

5.1 Introduction

Although the liquid/liquid interface would appear to be simpler than the solid/liquid one, investigations of that surface present some difficulty. When pure liquids are involved, the interface is homogeneous, and the orientation of the molecules of the two liquids on the contact surface is in doubt. Orientation of the molecules on both sides of the interface surface results in a region of oriented structure in the vicinity of that surface [1].

Consider two types of adsorption at the liquid/liquid interface:

(i) Substance soluble in one liquid (one liquid phase) only.

(ii) Substance soluble in both liquids (in both liquid phases).

5.2 Adsorption of a Substance Soluble in One Liquid

Adsorption, at the liquid/liquid interface, of a substance soluble in one liquid only is analogous to adsorption at the solid/liquid interface. Two liquid-phase systems in which water is one phase are of particular interest here. Very high interfacial tension is then observed especially when the other liquid is non-polar, such as a hydrocarbon. On the other hand, the properties of pure liquid water are fairly complex and not yet fully understood, and this applies especially at the interface. This situation is well illustrated by the variation of interfacial tension as a function of log c in the hexane–water + ethanol system presented in Fig. 5.1a (the same effect is observed for methanol and propanol) [2]. At low alcohol concentrations in the aqueous phase we observe a sharp rise with concentration of interfacial tension at 283 K (curve *1*) and at 298 K (curve *2*); further

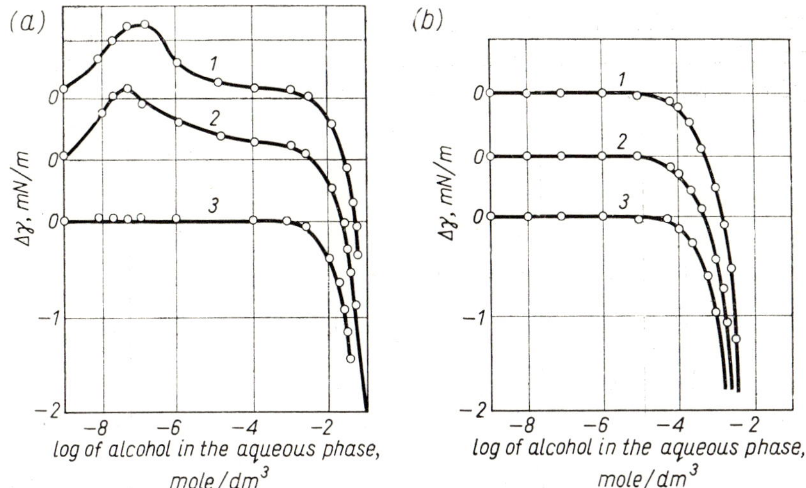

Fig. 5.1 Variation of the interfacial tension as a function of log c (after Grahame [2]): sa) in the hexane–water+ethanol system and (b) in the hexane–water+butanol (ystem at 283 K (*1*) and 298 K (*2*) and 313 K (*3*).

increase in concentration produces a lowering of that tension. At 313 K (curve *3*) we observe no anomalous transitional increase of the interfacial tension. When butanol is present in this system, there is no anomalous increase of interfacial tension at any of the given temperatures (Fig. 5.1b). Above this temperature range a change in water structure occurs, as evidenced by a sharp lowering of the interfacial tension at the water/hexane interface in the absence of dissolved substances. The anomalous increase of interfacial tension produced by alcohols is probably due to repulsion of their molecules by the stiff interfacial water layer, which can be broken only at higher alcohol concentrations. At higher temperatures the interfacial water layer is less stiff and the effect of the alcohols on the interfacial tension is normal.

Hutchinson calculated [3] the adsorption of higher fatty acids and aliphatic alcohols from benzene at the water/benzene interface (Fig. 5.2) on the basis of the Gibbs adsorption isotherm. His calculations involving the magnitude of the surface area occupied by one aliphatic alcohol molecule at the liquid/gas interface (about 0.2 nm^2) for complete surface coverage led him to conclude that the molecules are arranged vertically at that interface. The surface areas occupied by fatty acids proved, however, to be much larger (much lower adsorption isotherms). Calculations have shown that for such substances these areas are about 0.41 nm^2 per molecule. This indicates that the acid molecules are arranged parallel to the interface in the form of dimers (the molecules are linked by strong hydrogen bonds).

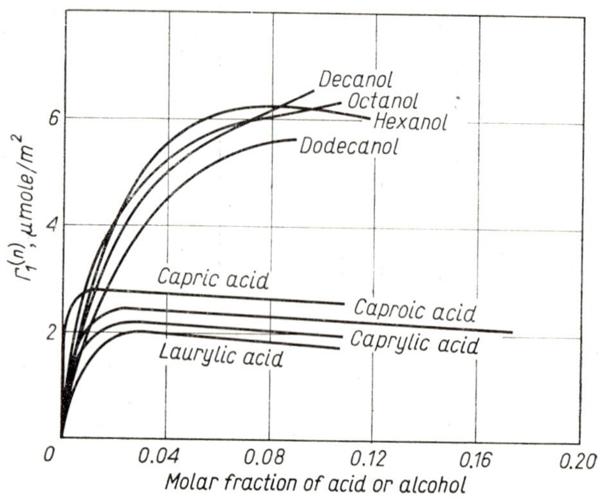

Fig. 5.2 Excess adsorption isotherms for higher fatty acids and aliphatic alcohols at the water/benzene interface (after Hutchinson [3]).

The effect of non-polar solvent structure on molecular orientation is well represented by the adsorption of caprylic acid from *n*-tetradecane or benzene on the water surface [3]. Adsorption of that acid is much greater for *n*-tetradecane than for benzene. Adsorption of caprylic acid may depend on the structure of the adsorption layer formed. In the case of *n*-tetradecane the aliphatic chains of the solvent and acid are arranged parallel to each other. The fairly strong interaction of the carboxylic groups of the acid with water results in vertical orientation both of the solvent and solute molecules. Benzene molecules, owing to their π electrons, are arranged flat on the water surface producing a parallel orientation of adsorbed fatty acid molecules forming dimers in the benzene solution.

Aliphatic alcohols show a weaker tendency to form dimers as compared with the strong hydrogen bonds formed by the acids with the water surface.

5.3 Adsorption of Substances Soluble in Both Liquids

Studies of the adsorption of polar molecules with a short hydrocarbon chain on the water/petroleum ether interface [4] have been interpreted using the Langmuir equation of state

$$\pi(A-A_0)=kT \tag{5.1}$$

where π is the surface pressure, k is a constant, A is the surface area occupied by one molecule at surface pressure π, and A_0 is the smallest surface area occupied by one molecule at the interface.

Figure 5.3 illustrates the $1/\pi$ vs. A plots for n-butyric acid and for n-butyl alcohol, for which A_0 is equal to 0.215 and 0.185 nm², respectively. It follows that the molecules being adsorbed are oriented vertically to the interface.

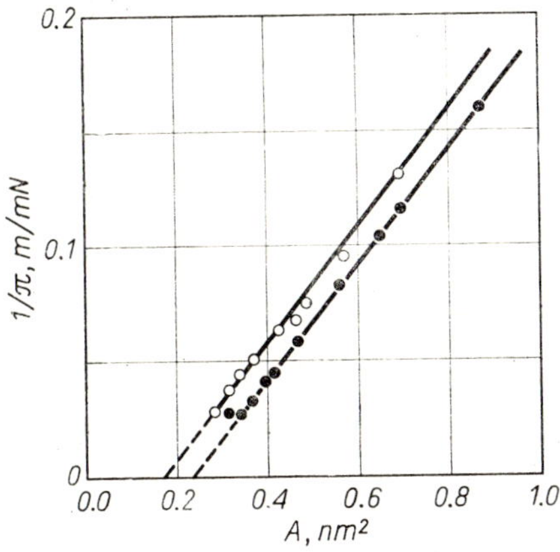

Fig. 5.3 Variation of $1/\pi$ with A for n-butyric acid (full circles) and n-butyl alcohol (open circles) at the water/petroleum ether interface (after Haydon and Taylor [4]).

The adsorption isotherm corresponding to equation (5.1) can be written in the form:

$$\frac{A_0}{A-A_0}\exp\frac{A_0}{A-A_0}=a\exp\left(\frac{\Delta_a G_m^s}{RT}\right) \tag{5.2}$$

where a is the activity of the substance dissolved in the water phase and $\Delta_a G_m^s$ is the molar Gibbs free energy of adsorption. This equation is used for the mobile adsorption layer.

Bartell [5] has shown that the adsorption of *n*-butyric acid and lower aliphatic alcohols are the same at the water/paraffin and water/air interfaces. This statement is based on a comparison of the interfacial tension and surface tension lowering at a given concentration of dissolved substance.

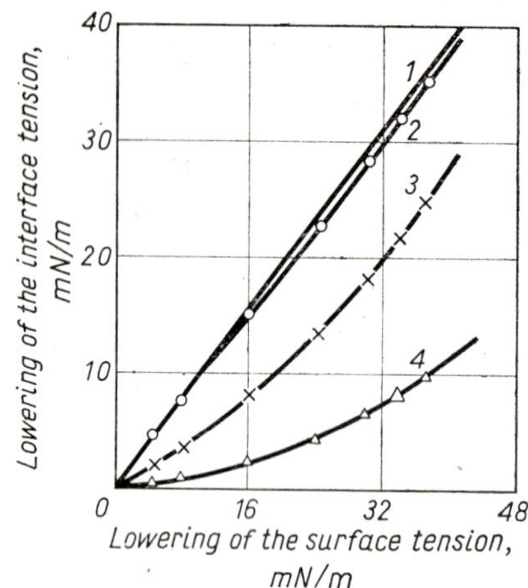

Fig. 5.4 The lowering of the interfacial tension by ethanol at the: *1* – water/air, *2* – water/*n*-heptane, *3* – water/benzene, and *4* – water/methyl–*n*-amyl ketone interfaces (after Bartell and Davis [5]).

Figure 5.4 presents the results of comparing the lowering of interfacial tension by ethanol at various interfaces the water/air system being used as reference. The adsorption of dissolved substance is smaller when the non-aqueous phase shows lower interfacial tension towards water, for instance, we have 34.3 $mN \cdot m^{-1}$ for benzene and 12.5 $mN \cdot m^{-1}$ for *n*-amyl ketone. The almost equal adsorption at the water/heptane (50.9 $mN \cdot m^{-1}$) and water/air interfaces is related to the Gibbs free energy of adsorption of the dissolved substance which, for the above interfaces, is practically constant.

References

[1] Franks, F. and Ives, D. J. G., *J. Chem. Soc.*, 741 (1960).
[2] Grahame, D. C., *J. Colloid Sci.*, **3**, 1725 (1955).
[3] Hutchinson, E., *J. Colloid Sci.*, **3**, 219, 235 (1948).
[4] Haydon, D. A. and Taylor, F. H., *Phil. Trans.*, **252**, 225 (1960).
[5] Bartell, F. E. and Davis, J. K., *J. Phys. Chem.*, **45**, 1321 (1941).

Chapter 6

The Nature of Molecular Interactions in Adsorption

6.1 Introduction

In previous chapters, we have discussed adsorption phenomena at various interfaces. Studies on the liquid/gas and liquid/liquid interfaces have been carried out by numerous authors over many years, and the molecular mechanism of adsorption has been explained fairly well for these cases.

The situation is quite different as regards the molecular mechanism of adsorption at the solid/gas and solid/liquid (solution) interfaces. There is as yet no definite answer to the question: what is the nature of interactions between molecules of the adsorbate and the adsorbent surface, i.e. what is the nature of the adsorption forces? Well-known adsorption theories developed at the beginning of the present century, and extended and modernized in the twenties and thirties have contributed little to our understanding of molecular interactions in adsorption. This situation is mainly the result of a lack of detailed knowledge of the methods of obtaining adsorbents with an accurately determined geometrical structure and a definite chemical surface character. Only in the last two decades have a number of works appeared which have thrown considerable light on this problem. This contributions have permitted an expression of the adsorptivity of many solids in the form of physicochemical quantities dependent solely on adsorbent–adsorbate interactions. One should mention here the works of Kiselev, Everett, Giles, Snyder and others.

Adsorption interactions are distinguished from molecular interactions in gases by the fact that distances between adsorbate molecules and the adsorbent surface (ions, atoms or molecules on that surface) are small compared with the distances between gas molecules. Thus adsorbent–adsorbate interactions are analogous to molecular interactions in condensed media (e.g. solution). Adsorption phenomena have, therefore, much in common with association or solvation in liquids.

Investigation of adsorption forces is of great importance not only for the physical chemistry of surface phenomena, but also for the experimental testing of a theory of interactions of molecules of more complex structure, especially in solution.

We shall primarily discuss here molecular interactions occurring in physical adsorption, whereas the forces controlling chemisorption will be dealt with briefly only at the end of the chapter.

6.2 Types of Molecular Interactions in Adsorption

The chemical character of the adsorbent surface determines the nature and energy of interaction between the adsorbate molecules (more generally, the molecules of the surface layer) and the solid.

The nature of molecular interactions is in principle quantum-mechanical. However, there is as yet no complete theory for such interactions and no general expression for such interaction potentials at small distances. That potential is thus expressed as the sum of independent contributions: dispersion, electrostatic and chemical interactions.

The different types of interaction which occur depend on the type of adsorbate molecules and the nature of the adsorbent surface. Kiselev [1, 2], Everett [3], and Barrer [4] have classified these interactions qualitatively as non-specific or specific.

The non-specific interaction is a general interaction, occurring between any species. This is primarily a dispersion interaction related to the co-ordinated motion of electrons in the molecule.

The specific interaction is due to a special distribution of the external (peripheral) electron density in molecules, and is related to local concentrations of negative and positive electric charges within a molecule. However, the specific interaction is not a classical electrostatic one, and only in cases of larger distances between molecules does it reduce to the electrostatic one. The hydrogen bond is an example of such a specific molecular interaction. A still stronger interaction, for instance that involving complete charge transfer between the partners of a donor–acceptor co-ordination bond, leads to the loss of chemical individuality of the interacting molecules.

6.3 Classification of Molecules and Adsorbents According to their Molecular Interactions

6.3.1 Classification of Molecules

Attempts to classify substances according to the ability of their molecules to interact have been made by many authors. Ewell, Harrison and Berg proposed a classification based on the ability of molecules to form hydrogen bonds [5]. They used their classification primarily to characterize solvents (Table 6.1). Pimentel and McClellan [6] supplemented that classification

Table 6.1 The classification of solvents after Ewell *et al.* [5], and Pimentel and McClellan [6]

Class	General characteristic	Solvents
AB*	Ability to form a cubic lattice of strong hydrogen bonds	water, glycol, glycerine, aminoalcohols, hydroxylamines, hydroxy acids, nitrophenols, amides
AB	Presence of active hydrogen atoms as well as of oxygen, nitrogen and fluorine atoms showing electron-donor properties. Weaker hydrogen bonds	alcohols, acids, phenols, prim. and sec. amines, oximes, ammonia, HF, HCN
B	Presence of atoms with electron-donor propeities (oxygen, nitrogen, fluorine). No active hydrogen atoms	ethers, ketones, aldehydes, esters, tert. amines (inclusive of pyridine derivatives); nitro compounds nitriles, olefins
A	Presence of active hydrogen atoms. Absence of electron donors	$CHCl_3$, CH_2Cl_2, CH_3CHCl_2, etc.
N	Lack of ability to form active hydrogen bonds	aliphatic hydrocarbons, mercaptans, chloroderivatives

with symbols based on the Lewis theory (B – basic, A – acidic, AB – amphoteric, and N – neutral).

Kiselev [1, 2, 7] developed his classification on a more general basis, i.e. on variations in the distribution of peripheral electron density of bonds and regions of the molecules. He distinguishes four groups of molecules: A, B, C and D.

Group A. This group includes molecules with spherically symmetric electron shells similar to noble gases; e.g. saturated hydrocarbon molecules (including only σ-bonds).

Group B. In molecules of this group the negative charge concentrates on the peripheries of bonds or atom groupings, e.g. unsaturated and aromatic hydrocarbons, all molecules containing π-bonds (e.g. N_2), and functional groups possessing lone pairs of electrons (e.g. in ether, ketone, tert. amine, nitrile, sulphide, etc. molecules).

Group C. In these molecules a local concentration of positive charge occurs. The equivalent excess negative charge is uniformly distributed over the remainder of the molecule, e.g. organometallic compounds.

Group D. These molecules contain functional groups in which the positive charge concentrates in one part and the negative charge in another part, e.g. such functional groups as —OH, —NH_2, $\rangle$NH. In this group we classify water, alcohol, prim. and sec. amine, etc. molecules.

Of course molecules of many substances may be classified according to two of the given groups. This applies primarily to molecules with a large number of functional groups and π-type bonds.

The molecules of group A react only via the non-specific general interaction. A specific molecular interaction complements a non-specific one only when the interacting molecules can be classified in group B, C or D.

Comparison of the above two classifications of molecules reveals, of course, their similarity.

6.3.2 Classification of Adsorbents

The classification of adsorbents into polar and non-polar proves unsatisfactory when the nature of adsorption forces is considered. In recent years, adsorbents have been synthesized not only with a homogeneous geometrical surface structure but also with given chemical character. By introducing suitable functional groups on to the surface of a polar silica gel we can change its adsorption properties over a wide range. These properties may become analogous to those of non-polar adsorbents when the active OH groups at the gel surface are substituted by non-active groups e.g. OCH_3.

It is therefore more reasonable to classify adsorbents according to their surface chemical character, account being taken of the electronic charge distribution on their surface. According to Kiselev [1, 2, 7] one can distinguish three types of adsorbents.

Type I – non-specific adsorbents. At the surface there are no functional groups or exchangeable ions, e.g. graphitized carbon blacks, saturated hydrocarbons and polymers (e.g. polyethylene).

Type II – specific positive adsorbents. Adsorbents of this type have OH groups of acidic character at their surface, e.g. hydroxylated surfaces of acid oxides, especially $SiO_2 \cdot nH_2O$. The hydrogen atoms in surface OH groups are strongly protonated as a result of the presence of vacant *d*-orbitals on the silicon atom.

A similar charge distribution is also possible in the absence of acidic OH groups. In that case aprotonic acid groups or cations of small diameter should appear at the adsorbent surface. The compensating negative charge is then distributed over the internal bonds of the large complex anion, e.g. zeolites in which the positive charge of the surface exchangeable cations is compensated by the negative charge distributed within the large anions (AlO_4^-).

Type III – specific negative adsorbents. These adsorbents have on their surface bonds or groups of atoms with a negative charge concentrated on their peripheries. They are obtained most frequently by depositing on the surface of a non-specific adsorbent (i.e. adsorbent of type I), and in particular on the surface of graphitized carbon black, a monomolecular layer of molecules or macromolecules belonging to group B. These adsorbents are also obtained by introducing suitable functional groups, e.g. —CN, by suitable chemical modification of the surface of other adsorbents.

6.4 Molecular Interactions in Adsorption from the Gaseous Phase

6.4.1 Non-specific Interactions

Non-specific interactions occur in adsorption if type I non-specific adsorbents are used, when the adsorbed molecules may have any electronic structure. The non-specific interaction is due mainly to the occurrence of dispersion forces related to the co-ordinated motion of electrons in the reacting molecules and the geometrical arrangement of adsorbate molecules with respect to the adsorbent surface. At low surface coverage of type I adsorbents the mutual adsorbent–adsorbate interaction energy increases linearly with size of the adsorbed molecules for a homologous series.

A typical non-specific adsorbent with homogeneous surface and small specific surface area (6–30 m^2/g) is thermal black graphitized at about

3273 K. The particles of such carbon black are in the form of polyhedra whose planes are the exterior walls of the graphite crystal lattice [8, 9]. Figure 6.1 shows the variation of differential molar heats of adsorption (q_d) on such carbon black with the number of carbon atoms (n) in the molecules of the members of homologous series. We see that at low surface

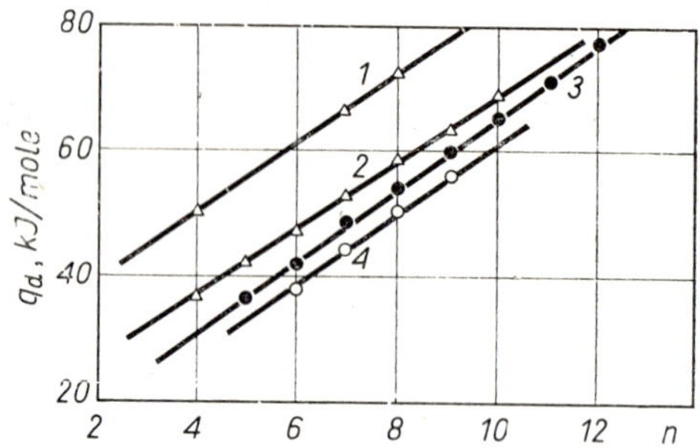

Fig. 6.1 Variation of the differential molar heats of adsorption with the number of carbon atoms (n) in members of homologous series (after Belakova *et al.* [10] and Kiselev *et al.* [11]). Adsorbent: graphitized carbon black; homologous series: *1*—*n*-acids, *2*—*n*-alcohols, *3*—*n*-alkanes, *4*—*n*-alkylbenzenes; temperature: 383 K.

coverage the differential molar heats of adsorption for derivatives of *n*-alkanes differ by a constant amount from molar heats of adsorption for the *n*-alkanes themselves with the same numbers of carbon atoms.

The differential molar heats of adsorption q_d for *n*-alkanes and their derivatives are additive quantities, and so $q_{d(CH_3)}=8.79$, $q_{d(CH_2)}=6.69$, $q_{d(OH)}=8.79$, $q_{d(ethers)}=5.44$, $q_{d(COOH)}=25.94$ kJ/mole (in the temperature range 323–473 K).

At higher coverages of the graphitized carbon black, mutual interactions frequently occur between the adsorbate molecules. This leads to an increase in the differential heat of adsorption. A particularly high increase of that heat occurs when the adsorbed molecules can form hydrogen bonds with each other (e.g. molecules classified in group D).

Non-specific interaction also occur when adsorbents of types II or III are used and the adsorbed molecules belong to group A (groups N according to Pimentel and McClellan). Specific interactions may also occur on the surface of the adsorbents. This is not possible in the case of non-polar adsorbate molecules (the induction interaction energy is small). It has been shown that this is eventually the case for adsorption of *n*-hexane and CCl_4 on the surface of silica gels whose OH groups have been removed

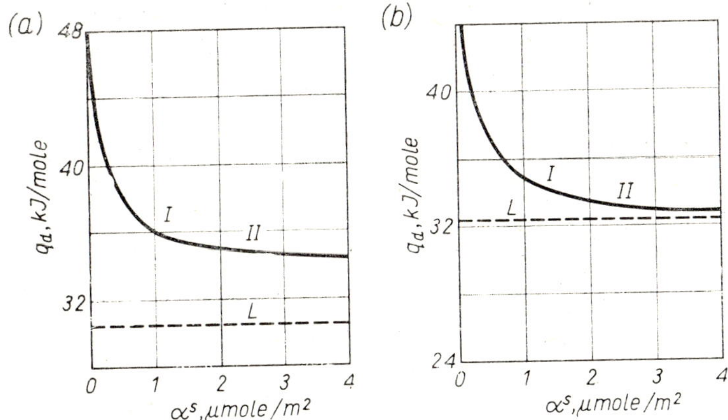

Fig. 6.2 Variation of the differential molar heats of adsorption of *n*-hexane (a) and CCl_4 (b) with coverage of the adsorbent surface (***L*** being the heat of condensation): silica gel hydroxylated (*I*) and deprived of hydroxyl groups (*II*) (after Kiselev *et al.* [12–15]).

as a result of heating under high pressure. The results of measurement of differential heat of adsorption are shown in Fig. 6.2. From the plots it is seen that the interaction energies between the silica gel with hydroxylated surface (*I*) and that devoid of OH groups (*II*) and the adsorbate molecules with symmetrical electron structure, possessing only σ-bonds, are practically equal.

6.4.2 Specific Interactions

In general specific molecular interactions may be affected by induction or orientation forces, hydrogen bonds, formation of molecular complexes, or by atomic or ionic bonds. Specific interaction occurs with type II or III adsorbent and adsorbate molecules with appropriate functional groups or π-electron molecules belonging to the Kiselev B, C or D groups or else to the Pimentel and McClellan A, B and AB groups.

6.4.2.1 *Specific Interactions on Type II Adsorbents*

Type II specific adsorbents are most common in practice. Typical representatives are: silica gels, aluminium oxides as well as zeolites with their cationized surfaces. When discussing molecular interactions in adsorption on these adsorbents, the chemical structure of their surface should asso be considered.

Silica gel consists of globules composed, in a disordered manner, of SiO_4 tetrahedra (Fig. 6.3). These globules are formed in the course of condensation of orthosilicic acid; the OH groups on the surface of the globules do not react. The adsorption properties of silica gel depend on the mutual orientation of the edges of the surface SiO_4 tetrahedra to which the hydroxyl groups are bound.

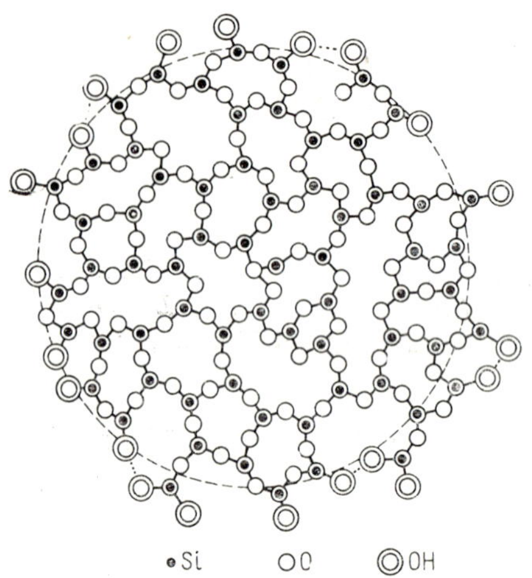

Fig. 6.3 Diagram of the globules forming the skeleton of silica gel (after Kiselev *et al.* [16]).

Owing to the disordered arrangement of surface SiO_4 tetrahedra, the surface OH groups bonded to silicon atoms (⟩Si—OH, silanol groups) are not equivalent as regards their adsorptivity. Although siloxal groups (⟩Si—O—Si⟨) also occur on the silica gel surface, nevertheless the specific adsorptive interaction depends on the number and distribution of OH groups per unit area of adsorbent (α^{s}_{OH}).

Studies of Kiselev *et al.* [17–20], Snyder and Ward [21], McDonald [22], Egorov *et al.* [23], Cusumano and Low [24–26] and other authors [27–29] have shown that there are, in principle, two kinds of hydroxyl groups on the surface of silica gel:

(i) Free (isolated) OH groups – active centres of A type; here the distance to the nearest OH groups is 0.50–0.52 nm.

(ii) Bonded OH groups – active centres of B type, belonging to neighbouring silicon atoms; here the distance between the nearest OH groups is 0.25–0.26 nm, so they can interact via hydrogen bonds (Fig. 6.4).

The total number of surface OH groups on silica gel can be determined by chemisorption of HMDS ($(CH_3)_3Si—NH—Si(CH_3)_3$). Davydov *et al.* [30] found that in the reaction with $ClSi(CH_3)_3$ in principle only free hydroxyl groups take part. Snyder and Ward [21] have shown that reactions of silanes with hydroxyl groups proceed in two steps. In the first step the

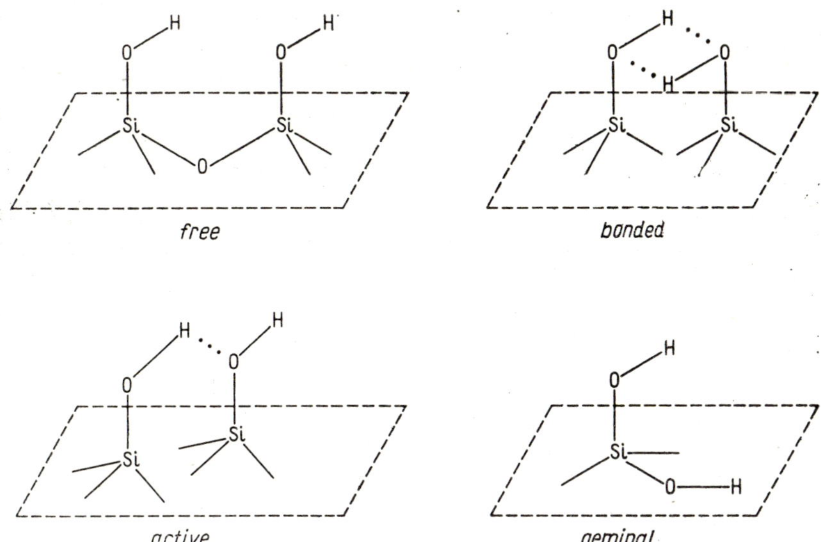

Fig. 6.4 Types of hydroxyl groups occurring at the surface of silica gel. The bonded and active OH groups occur when the crystal lattice is deformed or the surface bent.

reaction is very rapid, for which active adsorption centres other than free hydroxyl groups must be responsible. These centres have been referred to as active OH groups. Snyder and Ward found that such adsorption centres occur in some kinds of B type centres, in which one of the two protons of the bonded hydroxyl groups does not take part in the hydrogen bond because of their spatial arrangement. The activity of such bonded OH groups is therefore greater than that of the free groups. Reactive (active) OH groups are found to be stable and more resistant to the process of silica gel dehydration than other bonded OH groups.

Snyder [31] has distinguished a fourth type of hydroxyl group bonded to silicon (geminal OH groups).

It has been repeatedly stated [20, 30, 32, 33] that the surface concentration of OH groups in hydrated silica gels depends on the temperature of treatment of the gel and not on the specific adsorbent surface area. For wide-porous silica gels, after extensive evacuation at 423–473 K, the surface concentration of hydroxyl groups, α^s_{OH}, is approximately constant [20] and equals about 4.6 OH groups per 1 nm^2 of surface area [29, 30]. Armistead *et al.* [28] have found that about 1.4±0.1 of the surface OH groups (1.4 adsorption centres of A type) per 1 nm^2 are free OH groups, and 3.2 OH groups (1.6±0.1 adsorption centres of B type) per 1 nm^2 are bonded groups.

The number of free and active hydroxyl groups on the surface of silica gel is closely related to pore dimensions. Wide-porous silica gels contain more free OH groups, whereas active and bonded OH groups predominate on the surface of narrow-porous silica gels. Figure 6.5 shows the dependence of the relative concentration of active hydroxyl groups, α^s_a/α^s_t, (α^s_a is the concentration of active groups, α^s_t is the concentration of all surface OH groups), on pore diameter [21, 34].

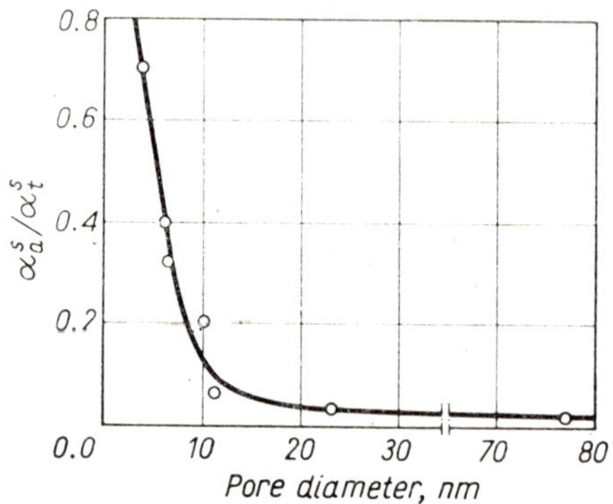

Fig. 6.5 Variation of the relative concentration, α^s_a/α^s_t, of active hydroxyl groups with the microporous structure of silica gels (after Snyder *et al.* [21, 34]).

As was shown by Peri and Hensley [35], complete hydration of silica gels yields a concentration of about 7.9 OH group per 1 nm^2. However, geminal OH groups are interlinked by hydrogen bonds to hydroxyl groups of neighbouring silicon atoms.

Polyfunctional adsorbate molecules as well as those possessing double or π-bonds are most strongly adsorbed on active hydroxyl groups, whereas the monofunctional molecules are adsorbed with equal strength on active and free OH groups [36]. In general we conclude that the adsorption energy of surface hydroxyl groups increases in the order: bonded groups $<$free groups$<$active groups.

Exhaustive spectral and calorimetric investigations of silica gel surface OH group interactions with various adsorbate molecules have been made by Kiselev [19] and Curthoys *et al.* [37].

Molecules with only one strongly adsorbing group react with the free or active hydroxyl groups of silica gel – localized adsorption. If, instead, the adsorbate molecules have several functional groups, then, depending on the structure of the molecule and type of functional groups present, two positions are possible relative to the nearest active sites on the adsorbent surface [30]:

(i) when no functional group dominates the interaction energy with the adsorbent active sites, every group can occupy an optimal position with respect to the nearest OH group – semilocalized adsorption;

(ii) when the molecule contains several functional groups reacting with different energies with the OH groups, then only one of these groups is localized. The remaining groups have a very limited possibility of assuming the most advantageous position with respect to the nearest active sites of the adsorbent.

Ultra-violet or infra-red spectra of the adsorbed substances reveal certain molecular deformations. This shows that for specific adsorption interactions with protonated hydrogen of the OH groups, electron transfers occur to different extents in the molecule being adsorbed. The hydrogen bond or, in the case of unsaturated and aromatic hydrocarbons, the formation of π-complexes may often be responsible for such specific interaction. For instance, when group D molecules (water, alcohols, etc.) are adsorbed on the hydroxylated silica gel surface, hydrogen bonds are formed according to the scheme presented in Fig. 6.6.

Since surface hydroxyl groups are responsible for specific adsorption on silica gel, the surface of these gels used as adsorbents should be hydrated to a maximum degree. Activation of these adsorbents at 150–200°C is quite sufficient, since most water molecules are then desorbed without affecting the number of surface OH groups [38, 39]. Higher activation temperatures reduce the activity of the silica gel due to the removal of surface hydroxyl groups (Fig. 6.7). Activation at temperatures 200–400°C causes condensation of most of the bonded (and geminal) OH groups according to Fig. 6.7a. At about 400° Cthe free hydroxyl groups start shifting on the surface and form active groups (Fig. 6.7b). Above that temperature the active groups undergo condensation (arrow pointing upwards in Fig. 6.7b). As a result of condensation, the adsorbent molecules also combine with liberation of water and the surface area decreases (Fig. 6.7c).

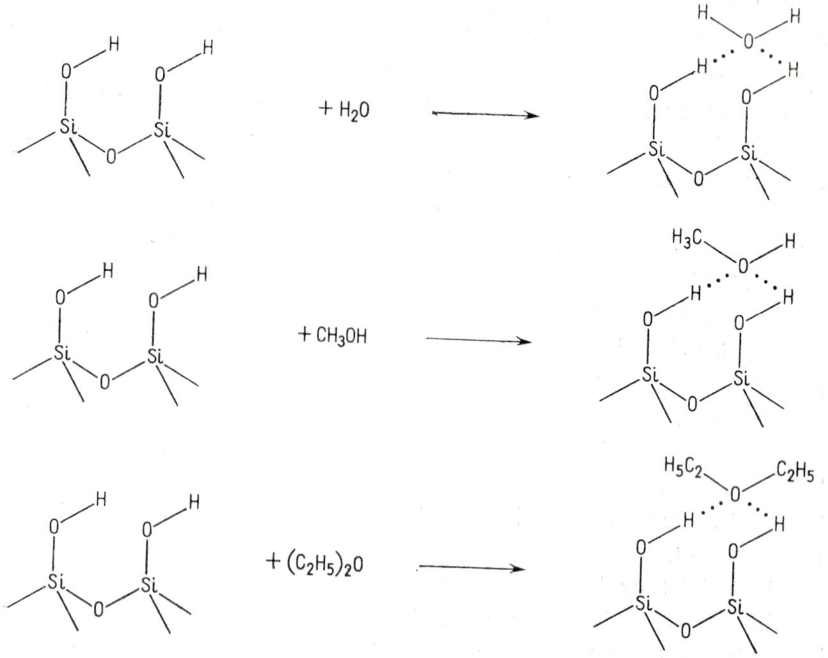

Fig. 6.6 Adsorption of D-type (according to Kiselev [38]) molecules on the hydroxylated silica gel surface.

At sufficiently high temperatures, silica gel is completely dehydrated, thus depriving it of its ability to adsorb unsaturated and polar adsorbate molecules.

Positive specific adsorbents include those with cationized surfaces such as zeolites (cf. Chapters 3 and 4); the walls of their channels are built of SiO_4 and AlO_4 tetrahedra. The negative charge is distributed over all the Al—O bonds inside the large complex anion and is compensated by the positive charge of cations which protrude on the zeolite surface. Such a surface structure is similar to the hydroxylated one with protonized hydrogen. When molecules containing atoms with free electron pairs or π-type bonds are adsorbed, we expect a qualitative analogy between their specific adsorption on the hydroxyl groups of the silica gel surface and on the cationized surface of zeolite channels.

Owing to the larger cation dimensions, classical Coulomb interactions may contribute largely to the specific interaction energy between the zeolite

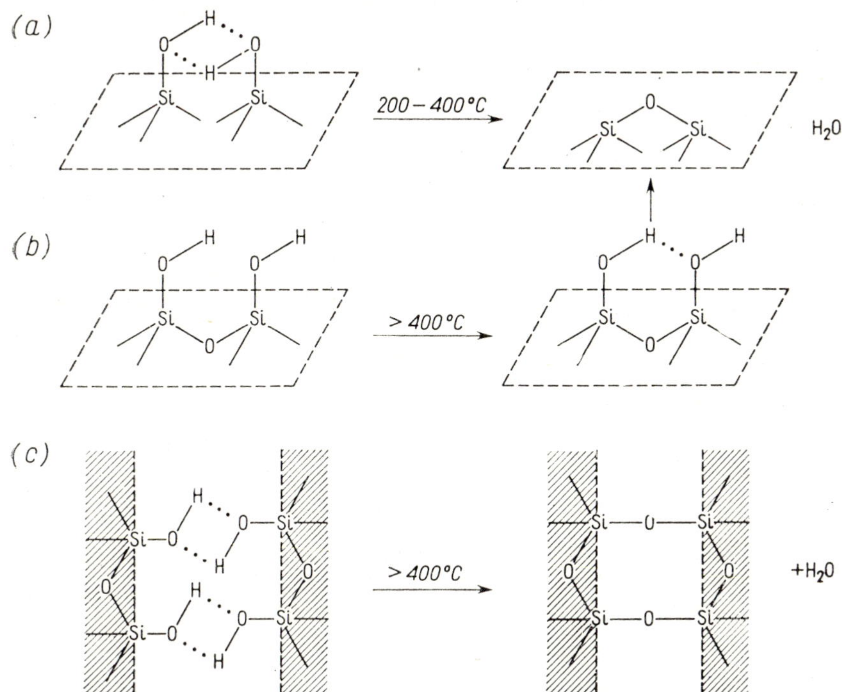

Fig. 6.7 Effect of temperature on the hydroxylated silica gel surface (after Snyder [31, 34]).

and such adsorbate molecules. Strong non-specific interaction of adsorbate molecules also takes place in narrow zeolite channels.

For such adsorbents we simultaneously encounter strong non-specific and specific interactions of the zeolite channel walls with adsorbate molecules. As a result, their bonds with surface cations resemble hydrogen bonds. Evidence for this specific interaction is found by comparing the variation of differential heats of adsorption of ethyl ether and *n*-pentane on graphitized carbon black, hydroxylated silica gel and X-type zeolite at concentration α^s (Fig. 6.8).

Type II adsorbents (specific, positive) also include aluminium oxide. The type of oxide most frequently used is γ-Al_2O_3 of specific surface area 100–200 m^2/g.

Peri [40], based on his own studies and on the suggestion of Lippens [41], developed a model of the γ-Al_2O_3 surface layer structure (see Fig. 3.38, p. 92), according to which the upper surface layer is built of O^{2-}

oxygen ions, the next lower layer consisting of Al^{3-} aluminium ions. Only about 3/4 of the sites allocated to aluminium ions are occupied, while the remaining sites exist in equilibrium as surface defects.

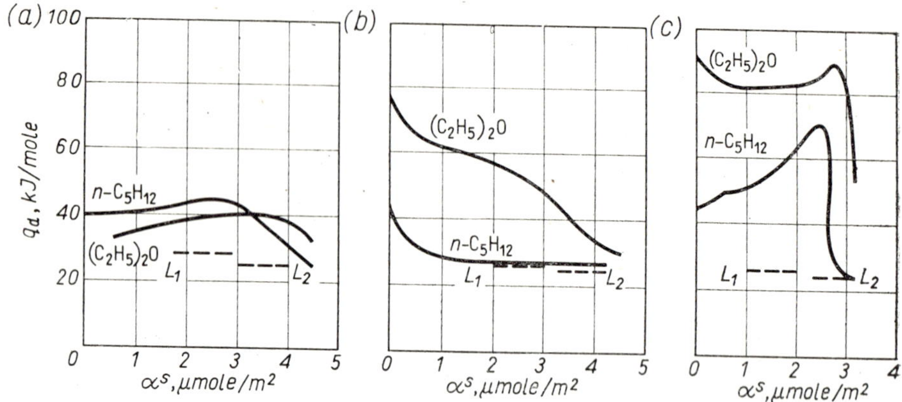

Fig. 6.8 Variation of the differential molar heat of adsorption of ethyl ether and *n*-pentane vapours with surface concentration α^s for graphitized carbon black (a), silica gel with hydroxylated surface (b) X-type zeolite (c) (after Kiselev *et al.* [10–13]).

In recent years, exhaustive investigations of the aluminium oxide surface structure have been conducted by Peri [40, 42–44], based on gravimetric experimental data and infra-red spectroscopy. Peri's model of the aluminium oxide surface is illustrated in Fig. 6.9. At the surface of the dehydrated aluminium oxide we have oxygen ions (Fig. 6.9a). At low temperatures in the presence of water vapour a completed monolayer of OH^- ions can be formed on the Al_2O_3 surface (Fig. 6.9b). In the course of dehydration the neighbouring OH^- ions combine, but only 2/3 of these ions can be removed as water vapour without affecting the local ordering. Further dehydration causes a disordered distribution of oxygen ions on the Al_2O_3 surface. After all OH^- ion pairs are removed, we obtain the surface shown in Fig. 6.10. The remaining hydroxyl ions cover about 10% of the aluminium oxide surface; they form, with the neighbouring oxygen ions, five types of active centres, A to E, as shown in Fig. 6.10, type A being most basic and type C most acidic.

Flockhardt *et al.* [45, 46] have postulated the existence of two types of defect on the surface of aluminium oxide heated to temperatures exceeding 773 K (see Fig. 6.11). The first type occurs when there is a deficiency of two or more oxygen ions in the surface layer. At such sites aluminium ions are

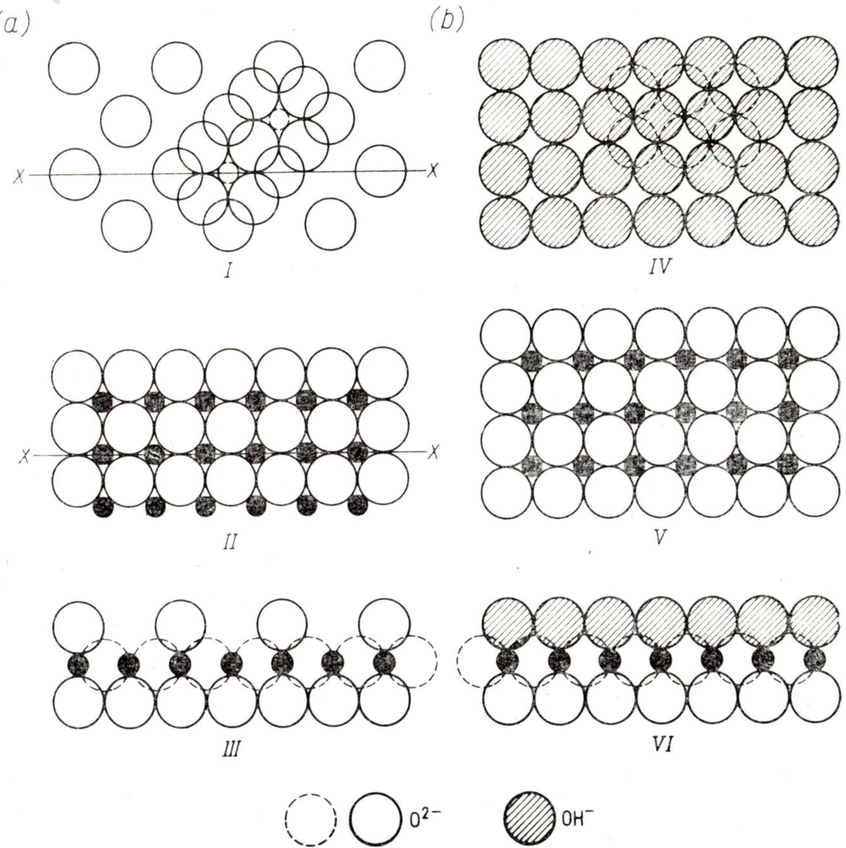

Fig. 6.9 Model of the ideal surface of aluminium oxide of spinel type proposed by Peri [40, 42–44]: (a) surface of dehydrated aluminium oxide: *I*–surface lattice of oxygen ions after the water has been removed, *II*–the next lower layer, *III*–vertical cross-section of the surface; (b) hydrated surface of aluminium oxide: *IV*–surface layer of OH^- ions, *V*–the next lower layer, *VI*–vertical cross-section of the surface.

exposed in an anomalous way on the surface (Fig. 6.11a). Owing to the accumulation of positive charges such a surface site becomes an electron acceptor (Lewis acid).

The second type is due to the occupation of a surface site by two or more oxygen ions (Fig. 6.11b). Such defects are of the electron donor type (Lewis base).

In many cases aluminium ions may be replaced by hydrogen ions, when proton surface defects occur.

When water is added to aluminium oxide, a chemisorbed layer is formed in which every water molecule is bonded to two surface oxygen atoms. At 573 K much of the adsorbed water is removed, and the remaining water molecules react with the surface, forming surface-ionized OH groups (ions).

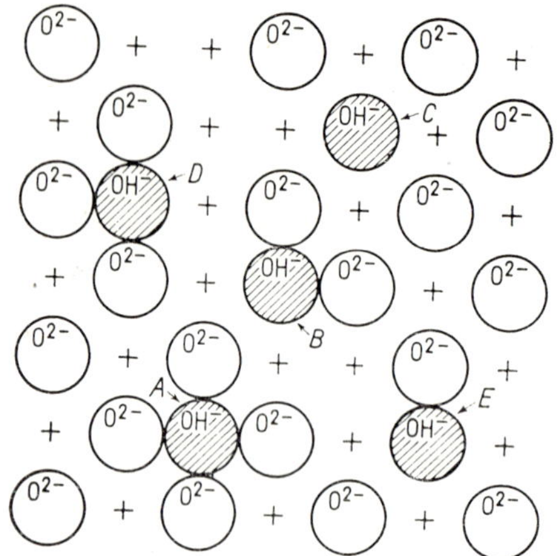

Fig. 6.10 Model of aluminium oxide surface after strong dehydration (after Peri [40,42–44]).

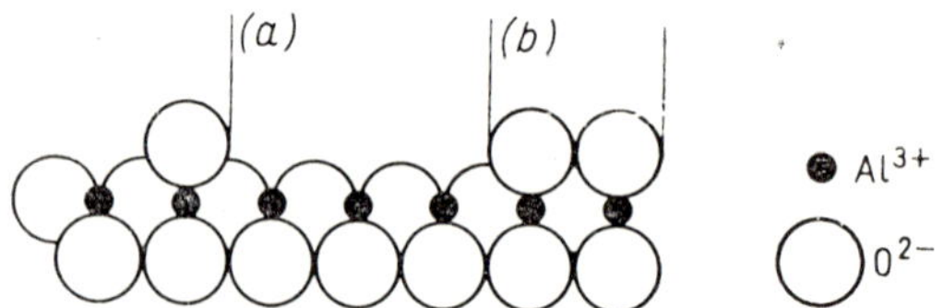

Fig. 6.11 Surface defects of aluminium oxide heated at 773 K: (a) electron-acceptor defects, (b) electron-donor defects (after Flockhardt *et al.* [45, 46]).

Aluminium oxide heated at 673 K under vacuum still has about 6 OH^- ions per 1 nm^2 of its surface area. Further heating to 1073 K and higher temperatures causes the removal of all hydroxyl ions from the aluminium oxide surface. At temperatures above 1273 K an inactive α-Al_2O_3 form of aluminium oxide is obtained with a specific surface area of 1 m^2/g [47].

From these studies, four types of active sites occur on the Al_2O_3 surface:

(i) Al^{3+} ions (acidic),
(ii) O^{2-} ions (basic),
(iii) ionized hydroxyl groups (basic),
(iv) proton defects (electron-acceptors).

Most substances are adsorbed on Al^{3+} ions around which a strong electric field exists; molecules liable to polarization (unsaturated and aromatic hydrocarbons) are adsorbed particularly strongly. Their weak localization on aluminium oxide results in greater selectivity of adsorption of various aromatic hydrocarbons. The weak localization of such substances on aluminium oxide is due to the structure of its surface layer, where strong and weak (with respect to the given adsorbate) sites are arranged in regular arrays on the surface (see Fig. 6.9). Molecules which do not have localized groups are thus adsorbed along rows of active sites on the surface of aluminium oxide, interacting with all the nearest sites. Linear molecules are therefore more strongly localized on the surface of aluminium oxide than non-linear molecules.

Molecules with one or more acidic groups are adsorbed on sites of basic character, since the acidic groups are strongly localized on these sites. Such molecules are adsorbed on surface proton defects just as on acidic sites.

The hydroxyl groups on the surface of aluminium oxide play a minor role in adsorption, as is confirmed by the fact that adsorbing properties of aluminium oxide increase with growth of activation temperature from 273 to 1273 K, whereas at 1073 K we observe a complete lack of surface hydroxyl groups.

6.4.2.2 *Specific Interactions on Type III Adsorbents*

Our knowledge of specific negative adsorbents, which can be obtained by chemical or adsorptive modification of the adsorbent surface, is limited. Kiselev [7, 33] has described a specific negative adsorbent obtained from a wide-porous silica gel by chemical modification of its surface using organosilicone compounds containing nitrile groups. He thus obtained an adsorbent whose surface was covered with CN instead of OH groups. The presence on the surface of such an adsorbent of CN groups with electrons concentrated on their peripheries results in strong specific adsorption with D type adsorbate molecules (formation of hydrogen bonds). Because of the high dipole moment of the CN group, molecules with high dipole moments (which, however, cannot form hydrogen bonds with those groups) strongly react with an adsorbent surface covered with CN groups (e.g. nitriles and nitro-derivatives; i.e. molecules belonging to group B). On such surfaces, group C molecules ought to be strongly adsorbed. Kiselev *et al.* [48] have modified the adsorption behaviour of surface of Grafon carbon black by depositing on it a monolayer of polyethylene glycol. This surface is characteristic of type III adsorbents since it contains $-CH_2CH_2O-$ groups with ether oxygen atoms with two free electron pairs. Such an adsorbent strongly reacts specifically with D group molecules. Its interaction with nitro-derivative and nitrile molecules is weaker since the dipole moments of the polyethylene glycol ether groups is much smaller than that of the CN group of the chemically modified wide-porous silica gel.

6.5 Molecular Interactions in Adsorption from Solution

In adsorption on the solid/gas (vapour) interface at low surface coverage (at low adsorbate pressure) the mechanism depends solely on the nature of the forces acting between adsorbent and adsorbate molecules (Fig. 6.12). The molecular mechanism of adsorption of a given substance from solution on a given adsorbent is much more complicated. In the case of the simplest

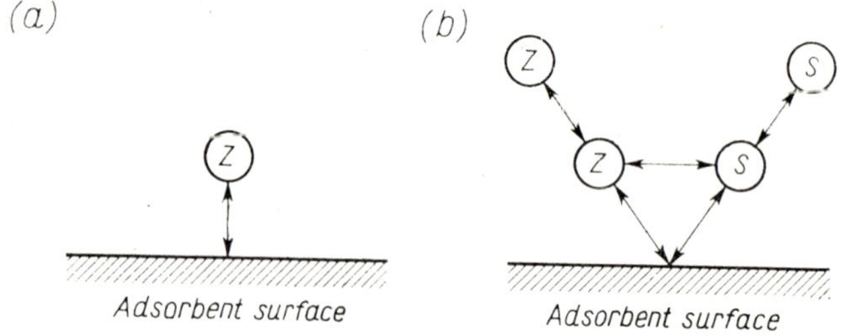

Fig. 6.12 Scheme of molecular interactions occurring in adsorption: (a) from the gaseous phase, (b) from a binary solution. Z – adsorbate molecule, S – solvent molecule.

(binary) solution adsorption depends, even at low concentrations, on the following factors (Fig. 6.12):

(i) forces acting between the adsorbate molecules (Z) and the adsorbent surface,

(ii) forces acting between the solvent molecules (S) and the adsorbent surface,

(iii) forces acting between the molecules of the solution components (Z and S) both in the surface layer and in the bulk phase.

Investigations of the molecular mechanism of adsorption on the solid/solution interface encounter considerable difficulties because of the complexity of the system. Among the methods used, liquid adsorption chromatography deserves mention. Snyder [31], Ościk *et al.* [49–53], Soczewiński and Gołkiewicz [54, 55] and others have shown that molecular interactions can be estimated from relationships between certain chromatographic parameters and the type of adsorbent, mobile phases and substances used. The relationship between chromatographic quantities (e.g. R_M) and thermodynamic adsorption functions allows us to estimate adsorption equilibrium constants and the magnitude of molecular interactions [49, 50]. On the other hand, if we assume a simple molecular model of adsorption chromatography, it is possible to determine the way adsorbed molecules react with the adsorbent and how they are oriented on the adsorbent surface [54, 55].

It must be emphasized that the solvent often has a decisive effect on adsorption from solution. Interactions between the adsorbent and adsorbate molecules during adsorption from the gaseous phase applies also to solvent molecules. Strong specific interactions of solvent molecules with the adsorbent surface block its active sites and considerably lower the adsorption of the dissolved substance. Also, strong molecular interactions between solution components in the bulk phase generally have a significant negative effect on their interaction with the adsorbent surface. Thus high solubility is usually correlated with weak adsorption.

It should be emphasized that molecular interactions in condensed systems (and thus in the solid/liquid system) differ from interactions in the gaseous phase in view of the much smaller distances between the interacting molecules.

The complex molecular mechanism of adsorption from solution is still far from being fully understood.

6.6 Nature of Adsorption Forces in Chemisorption

We can distinguish, as mentioned in Chapter 3, two types of adsorption: physical adsorption and chemical adsorption (chemisorption), between the adsorbent surface and adsorbate molecules.

So far we have only considered physical adsorption. Chemical adsorption occurs when we have chemical binding of adsorbate molecules with the adsorbent surface. As in every chemical compound, chemisorptive bonds involve covalent forces with a greater or smaller contribution from ionic forces.

In general, we can say that in chemisorption, the adsorbed molecules and adsorbent form a homogeneous system whereas, in physical adsorption, they can be considered as two independent systems.

References

[1] Kiselev, A. V., *Zh. Fiz. Khim.*, **38**, 2753 (1964).
[2] Kiselev, A. V., *Disc. Faraday Soc.*, **40**, 205 (1965).
[3] Everett, D. H., in: *Gas Chromatography 1964*, A. Goldup (Ed.), London 1965, p. 219.
[4] Barrer, R. M., *J. Colloid and Interface Sci.*, **21**, 415 (1966).
[5] Ewell, R. H., Harrison, J. M. and Berg, L., *Ind. and Eng. Chem.*, **36**, 871 (1944).
[6] Pimentel, G. C. and McClellan, A. L., *The Hydrogen Bond*, W. H. Freeman Co., New York–San Francisco 1960.
[7] Kiselev, A. V., in: *Gas Chromatography 1964*, A. Goldup (Ed.), London 1965, p. 238.
[8] Popov, N. M., Kasatochkin, V. I. and Lukyanovich, V. M., *Dokl. Akad. Nauk SSSR*, **131**, 609 (1960).
[9] Graham, D. and Kay, W. S., *J. Colloid Sci.*, **16**, 182 (1961).
[10] Belakova, L. D., Kiselev, A. V. and Kovaleva, N. V., *Zh. Fiz. Khim.*, **40**, 1494 (1966).
[11] Kiselev, A. V., Migunova, I. A. and Yashin, Ya. I., *Zh. Fiz. Khim.*, **41**, 807 (1967).
[12] Belakova, L. D. and Kiselev, A. V., *Dokl. Akad. Nauk SSSR*, **119**, 298 (1958).
[13] Babkin, I. Yu. and Kiselev, A. V., *Zh. Fiz. Khim.*, **37**, 228 (1963).
[14] Aristov, B. G. and Kiselev, A. V., *Zh. Fiz. Khim.*, **38**, 1984 (1964).
[15] Bezus, A. G. and Kiselev, A. V., *Zh. Fiz. Khim.*, **40**, 580 (1966); Avgul, N. N., Kiselev, A. V. and Poshkus, D. P., *Adsorbtsya gazov i parov na odnorodnykh poverkhnostyakh* (*Adsorption of Gases and Vapours on Homogeneous Surfaces*), Izd. Khimya, Moscow 1975.
[16] Kiselev, A. V., Lygin, V. I., Neimark, I. E., Slinyakova, L. V. and Czen Wen Chen, *Koll. Zhurn.*, **20**, 52 (1958).
[17] Kiselev, A. V. and Lygin, V. I., *Koll. Zhurn.*, **21**, 581 (1959).
[18] Kiselev, A. V. and Lygin, V. I., *Usp. Khim.*, **31**, 351 (1962).
[19] Kiselev, A. V., *Disc. Faraday Soc.*, **52**, 14 (1971).
[20] Davydov, V. Ya., Zhuravlev, L. T. and Kiselev, A. V., *Zh. Fiz. Khim.*, **38**, 1518 (1964).
[21] Snyder, L. R. and Ward, J. W., *J. Phys. Chem.*, **70**, 3941 (1966).
[22] McDonald, R. S., *J. Phys. Chem.*, **62**, 1168 (1958).
[23] Egorov, M. M., Kvilividze, V. I., Kiselev, A. V. and Krasilnikov, K. G., *Kolloid Z. u. Z. Polymere*, **212**, 126 (1966).
[24] Cusumano, J. A. and Low, M. J. D., *J. Phys. Chem.*, **74**, 792, 1950 (1970).
[25] Cusumano, J. A., *J. Catalysis*, **23**, 214 (1971).
[26] Cusumano, J. A. and Low, M. J. D., *J. Colloid and Interface Sci.*, **38**, 245 (1972).
[27] Hair, M. L. and Hertl, W., *J. Phys. Chem.*, **73**, 4269 (1969).

[28] Armistead, C. G., Tyler, A. J., Hambleton, F. H., Mitchell, S. A. and Hockey, J. A., *J. Phys. Chem.*, **73**, 3947 (1969).
[29] Van Cauwelaert, F. H., Vermoortele, F. and Uytterhoeven, J. B., *Disc. Faraday Soc.*, **52**, 66 (1971).
[30] Davydov, V. Ya., Kiselev, A. V. and Zhuravlev, L. T., *Trans. Faraday Soc.*, **60**, 2254 (1964).
[31] Snyder, L. R., *Principles of Adsorption Chromatography*, Marcel Dekker Inc., New York–London 1968.
[32] Chertov, V. M., Dzhanbaeva, D. B., Plachinda, A. S. and Neimark, I. E., *Zh. Fiz. Khim.*, **40**, 520 (1966).
[33] Kiselev, A. V., Yashin, Ya. I., *Gazo-adsorbtsyonnaya khromatografiya* (*Gas Adsorption Chromatography*), Izd. Nauka, Moscow 1967.
[34] Snyder, L. R., *Separ. Sci.*, **1**, 191 (1966).
[35] Peri, J. B. and Hensley, A. L., Jr., *J. Phys. Chem.*, **8**, 2926 (1968).
[36] Snyder, L. R., *J. Chromatogr.*, **25**, 274 (1966).
[37] Curthoys, G., Davydov, V. Ya., Kiselev, A. V., Kiselev, S. A. and Kuznetsov, B. V., *J. Colloid and Interface Sci.*, **48**, 58 (1974).
[38] Kiselev, A. V., in: *Structure and Properties of Porous Materials*, D. H. Everett and F. S. Stone (Eds.), Butterworths Sci. Publ., London 1958, p. 195.
[39] Schultze, G. R. and Schmidt-Kuster, W. J., *Z. anal. Chem.*, **3**, 251 (1964).
[40] Peri, J. B., *J. Phys. Chem.*, **69**, 211 (1965).
[41] Lippens, B. C., Thesis, Delft University of Technology, Delft, The Netherlands, 1961.
[42] Peri, J. B. and Hannan, R. B., *J. Phys. Chem.*, **64**, 1526 (1960).
[43] Peri, J. B., *J. Phys. Chem.*, **69**, 211 (1965).
[44] Peri, J. B., *J. Phys. Chem.*, **69**, 220 (1965).
[45] Flockhardt, B. D., Scott, J. A. and Pinck, R. C., *Trans. Faraday Soc.*, **62**, 3 (1966).
[46] Flockhardt, B. D., Leith, I. R. and Pinck, R. C., *Trans. Faraday Soc.*, **65**, 542 (1969).
[47] Halpaap, H. and Reich, W., *J. Chromatogr.*, **33**, 70 (1968).
[48] Belakova, L. D., Kiselev, A. V., Kovaleva, N. V. and Rozenova, L. N., *Zh. Fiz. Khim.*, **42**, 2276 (1968).
[49] Ościk, J., *Przemysł Chem.*, **44**, 129 (1965).
[50] Ościk, J., Chojnacka, G. and Szczypa, B., *Przemysł Chem.*, **49**, 593 (1967).
[51] Ościk, J. and Różyło, J. K., *Chromatographia*, 4, 515 (1971).
[52] Ościk, J., *Fizicheskaya adsorbtsya iz mnogokomponentnykh faz* (*Physical Adsorption from Multicomponent Phases*), Izd. Nauka, Moscow 1972, p. 138.
[53] Ościk, J. and Chojnacka, G., *J. Chromatogr.*, **93**, 167 (1974).
[54] Soczewiński, E., *Analyt. Chem.*, **41**, 179 (1969).
[55] Soczewiński, E. and Gołkiewicz, W., *Chromatographia*, **4**, 501 (1971); **5**, 431, 594 (1972); **6**, 269 (1973).

Index

A

B

C

D

E

F

G

H

I

K

L

M

N

O

P

R

S

T

V

W

Z